An introduction to Hilbert space

An introduction to

Hilbert space

NICHOLAS YOUNG

Department of Mathematics
Glasgow University

CAMBRIDGE
UNIVERSITY PRESS

PUBLISHED BY THE PRESS SYNDICATE OF THE UNIVERSITY OF CAMBRIDGE
The Pitt Building, Trumpington Street, Cambridge, United Kingdom

CAMBRIDGE UNIVERSITY PRESS
The Edinburgh Building, Cambridge CB2 2RU, UK
40 West 20th Street, New York, NY 10011-4211, USA
477 Williamstown Road, Port Melbourne, VIC 3207, Australia
Ruiz de Alarcón 13, 28014, Madrid, Spain
Dock House, The Waterfront, Cape Town 8001, South Africa

http://www.cambridge.org

© Cambridge University Press 1988

First published 1988
Tenth printing 2005

Printed in The United Kingdom at the University Press, Cambridge.

British Library Cataloguing in Publication Data
Young, Nicholas
    An introduction to Hilbert space.
    1. Hilbert space
    I. Title
    515.7'33   QA322.4

Library of Congress Cataloguing in Publication data
Young, Nicholas
    An introduction to Hilbert space.

    Includes index.
    1. Hilbert space. I. Title.
    QA322.4.Y68   1988   515.7'33   87-24903

ISBN 0 521 33071 8 hardback
ISBN 0 521 33717 8 paperback

MCP

CONTENTS

FOREWORD

The basic notions of the theory of Hilbert space are current in many parts of pure and applied mathematics, and in physics, engineering and statistics. They are well worth a place in any honours mathematics course, and Chapters 1 to 8 of this book aim to present them in a way accessible to undergraduate students. A course in Hilbert space is likely to be the last analysis course for many students, and it should therefore be able to stand on its own: it should not depend for its motivation on further study of abstract analysis, but should as far as possible have a value which is apparent either on aesthetic grounds or for its scientific or practical applications. For this reason I have included more historical and background material than is customary, and have omitted some of the major theorems about Banach spaces which are traditionally taught in introductory courses on functional analysis, but which are really more appropriate to students who will be pursuing operator theory further (the closed graph, Hahn–Banach and uniform boundedness theorems). The second half of the book describes two substantial applications. One of these is standard: the Sturm–Liouville theory of eigenfunction expansions, and its role in the solution of the partial differential equations of mathematical physics by the method of separation of variables. The other (in Chapters 12 to 16) is less common, but is nevertheless ideal for a final year course. It is beautiful mathematics, it is relatively recent and visibly useful. It also entails the development of some standard operator theory along the way, and exhibits very well the connection between abstract analysis and the more classical field of complex analysis.

Although the book was written primarily for an undergraduate audience, I hope it may be found useful for graduate courses also. I firmly believe that functional analysis is best approached through a sound grounding in Hilbert space theory, and am confident that students will be better able to benefit from one of the many excellent advanced texts on functional analysis and its applications if they first master the material contained herein. Chapters 12 to 16 may also be of interest to some electrical engineers. Some recent developments, particularly in control and filter design, require familiarity with this aspect of operator theory.

Chapters 1 to 8 are based on a compulsory course of twenty lectures which I gave to third year honours students at Glasgow University, and the remainder of the book, with a few omissions, on an optional twenty lecture course for fourth year students. In forty lectures at the undergraduate level it should be possible to cover the whole book except for the Adamyan–Arov–Krein theorem and the proofs of Fatou's theorem and the existence of square roots of positive operators. Chapters 12 to 16 do not depend on Chapters 9 to 11: they can be read straight after Chapter 8.

The book presupposes introductory courses in real analysis, linear algebra and topology (metric spaces suffice). For Chapters 12 to 16, and some of the problems earlier in the book, elementary complex analysis is required. It is tacitly assumed in Chapters 9 to 11 that the reader has met differential equations before, though formal requirements are slight. I have taken pains *not* to assume knowledge of the Lebesgue integral: the reader is asked only to believe that there is a definition of integral which makes $L^2(a, b)$ complete and the continuous functions a dense subspace. However, I am obliged to admit that there are parts of Chapter 13 which will feel distinctly more comfortable to those who are familiar with Lebesgue measure.

I am grateful to Dr Frances Goldman and Dr Philip Spain for reading the text and making useful suggestions. I am also very thankful that, despite the all-conquering march of the word processor, Cambridge University Press was willing to accept manuscript, so that I do not have to thank anyone for his excellent typing.

INTRODUCTION

Functional analysis is a branch of mathematics which uses the intuitions and language of geometry in the study of functions. The classes of functions with the richest geometric structure are called Hilbert spaces, and the theory of these spaces is the core around which functional analysis has developed. One can begin the story of this development with Descartes' idea of algebraicizing geometry. The device of using co-ordinates to turn geometric questions into algebraic ones was so successful, for a wide but limited range of problems, that it dominated the thinking of mathematicians for well over a century. Only slowly, under the stimulus of mathematical physics, did the perception dawn that the correspondence between algebra and geometry could also be made to operate effectively in the reverse direction. It can be useful to represent a point in space by a triple of numbers, but it can also be advantageous, in dealing with triples of numbers, to think of them as the co-ordinates of points in space. This might be termed the geometrization of algebra: it enables new concepts and techniques to be derived from our intuition for the space we live in. It is regrettable that this intuition is limited to three spatial dimensions, but mathematicians have not allowed this circumstance to prevent them from using geometric terminology in handling n-tuples of numbers when $n > 3$. In the context of $\mathbb{R}^n$ one routinely speaks of points, spheres, hyperplanes and subspaces. Though such language comes to seem very natural to us, it still depends on analogy, and we must have recourse to algebra and analysis to verify that our analogies are valid and to determine which analogies are useful.

Once the geometric habit of mind was established in relation to $\mathbb{R}^n$ it was natural to extend it to other common objects of mathematics which enjoy a similar linear structure, such as functions and infinite sequences of real numbers. This is a bolder leap into the unknown, and we must expect that our intuition for physical space will prove a shakier guide than it was for $\mathbb{R}^n$. Indeed, the task of sorting out the right basic concepts in the geometry of infinite-dimensional spaces preoccupied leading analysts for some decades around the turn of the century. Thereafter the geometric viewpoint proved its worth, and came to provide the backdrop for the greater part of modern work in differential and integral equations, quantum mechanics and other disciplines to which mathematics is applied.

The study of differential and integral equations arising in physics was one of the main impulses to the emergence of functional analysis. A precursor of the subject can be seen in attempts by several mathematicians to treat such equations as limits in some sense of finite systems of equations. This approach had fair success, particularly in the hands of Hilbert, and it still has plenty of life in the domain of numerical analysis. Suppose, for example, one wishes to solve the integral equation

$$\int_0^1 K(x, y)f(y)\,\mathrm{d}y = g(x).$$

Here K and g are known continuous functions on $[0, 1] \times [0, 1]$ and $[0, 1]$ respectively, and one is looking for a continuous solution f. It seems natural to approximate this system by the finite system

$$\sum_{j=0}^{n-1} K\left(\frac{i}{n}, \frac{j}{n}\right)f_{jn} \cdot \frac{1}{n} = g\left(\frac{i}{n}\right),$$

$i = 0, 1, \ldots, n - 1$. Assuming that this system of n linear equations in the n unknowns $f_{0n}, \ldots, f_{n-1,n}$ has a unique solution, one might expect that, for large n, f_{jn} ought to be close to $f(j/n)$, at least under further conditions on K and g.

Hilbert was by no means the first to use this device. Fourier himself was led to introduce Fourier series in a rather similar way. In studying the conduction of heat he encountered the differential equation

$$\frac{\partial^2 V}{\partial x^2} + \frac{\partial^2 V}{\partial y^2} = 0,$$

subject to certain boundary conditions. By the method of the separation of variables he derived the solution

$$V(x, y) = \sum_{m=1}^{\infty} a_m\, e^{-(2m-1)x} \cos(2m - 1)\,y,$$

where the coefficients a_m are determined by the infinite system of linear equations

$$\sum_{1}^{\infty} a_m = 1,$$

$$\sum_{1}^{\infty} (2m-1)^2 a_m = 0,$$

$$\sum_{1}^{\infty} (2m-1)^4 a_m = 0,$$

$$\dots$$

Fourier handled these by taking the first k equations and truncating them to k terms. This gives a $k \times k$ system which has a solution $a_1^{(k)}, \dots, a_k^{(k)}$. On letting $k \to \infty$ Fourier obtained the desired solution of the infinite system.

Although this trick often worked, it has its dangers. Consider the infinite system

$$x_1 + x_2 + x_3 + \dots = 1,$$
$$x_2 + x_3 + \dots = 1,$$
$$x_3 + \dots = 1,$$
$$\dots$$

No choice of the x_j will satisfy this system, yet Fourier's limiting procedure would yield the apparent solution $x_j = 0$ for all j.

By virtue of powerful technique and a perception of what was important, Hilbert was able to make great contributions using this idea. Nevertheless, mathematicians came to regard the method as inadequate. It is clumsy and notationally complicated. The procedure of passage to the limit is difficult, and, indeed, it has been asserted that Hilbert did not always accomplish it correctly (see Reid, 1970). He himself did not arrive at the modern geometric viewpoint: Hilbert never used 'Hilbert space'. It was other mathematicians, particularly Erhardt Schmidt and Frigyes Riesz, who reflected on his results and discovered the right conceptual framework for them. Thereby they created a simpler, more elegant and more powerful theory. In this one does not try to reduce essentially infinite-dimensional questions to finite-dimensional geometry and then 'let $n \to \infty$': instead one develops the geometry of the objects of analysis as they naturally occur, using the familiar finite-dimensional geometry rather as a guide and analogy.

1

Inner product spaces

Some important metric notions such as length, angle and the energy of physical systems can be expressed in terms of the *inner product* (x, y) of vectors $x, y \in \mathbb{C}^n$. This is defined by

$$(x, y) = \sum_{i=1}^{n} x_i \bar{y}_i, \tag{1.1}$$

where $x = (x_1, \ldots, x_n)$, $y = (y_1, \ldots, y_n)$, and $\bar{y}_i$ is the complex conjugate of y_i. We wish to construct an infinite-dimensional version of this inner product. The most obvious attempt is to consider the space $\mathbb{C}^{\mathbb{N}}$ of all complex sequences indexed by $\mathbb{N}$. This is a complex vector space in a natural way, but it is not clear how we can extend the notion of inner product to it. If we replace the finite sum in (1.1) by an infinite one then the series will fail to converge for many pairs of sequences. We therefore restrict attention to a subspace of $\mathbb{C}^{\mathbb{N}}$.

1.1 Definition ℓ^2 denotes the vector space over $\mathbb{C}$ of all complex sequences $x = (x_n)_{n=1}^{\infty}$ which are square summable, that is, satisfy

$$\sum_{n=1}^{\infty} |x_n|^2 < \infty,$$

with componentwise addition and scalar multiplication, and with inner product given by

$$(x, y) = \sum_{n=1}^{\infty} x_n \bar{y}_n, \tag{1.2}$$

where $x = (x_n)$, $y = (y_n)$. $\qquad \square$

'Componentwise' means the following: if $x = (x_n)$, $y = (y_n) \in \ell^2$ and $\lambda \in \mathbb{C}$ then

$$x + y = (x_n + y_n)_{n=1}^{\infty},$$
$$\lambda x = (\lambda x_n)_{n=1}^{\infty}.$$

Let us check that this definition of inner product does make sense. Using the Cauchy–Schwarz inequality we find, for $k \in \mathbb{N}$,

$$\sum_{n=1}^{k} |x_n \bar{y}_n| = \sum_{n=1}^{k} |x_n||y_n|$$

$$\leqslant \left\{ \sum_{n=1}^{k} |x_n|^2 \right\}^{1/2} \left\{ \sum_{n=1}^{k} |y_n|^2 \right\}^{1/2}$$

$$\leqslant \left\{ \sum_{n=1}^{\infty} |x_n|^2 \right\}^{1/2} \left\{ \sum_{n=1}^{\infty} |y_n|^2 \right\}^{1/2}.$$

If (x_n) and (y_n) are square summable sequences then the latter expression is a finite number independent of k. Thus the series (1.2) converges absolutely, and so (x, y) is defined by (1.2) as a complex number for any $x, y \in \ell^2$.

It is obvious that ℓ^2 is closed under scalar multiplication but less so that it is closed under addition: we defer the proof of this to Exercise 1.12 below.

Let us make precise what it means to say that $\mathbb{C}^n$ and ℓ^2 are spaces with an inner product.

1.2 Definition   An *inner product* (or *scalar product*) on a complex vector space V is a mapping

$$(\cdot, \cdot): V \times V \to \mathbb{C}$$

such that, for all $x, y, z \in V$ and all $\lambda \in \mathbb{C}$,

  (i) $(x, y) = (y, x)^-$;

  (ii) $(\lambda x, y) = \lambda(x, y)$;

  (iii) $(x + y, z) = (x, z) + (y, z)$;

  (iv) $(x, x) > 0$ when $x \neq 0$.

An *inner product space* (or *pre-Hilbert space*) is a pair $(V, (\cdot, \cdot))$ where V is a complex vector space and $(\cdot, \cdot)$ is an inner product on V. $\square$

It is routine to check that the formulae (1.1) and (1.2) do define inner products on $\mathbb{C}^n$ and ℓ^2 in the sense of Definition 1.2. There are many other inner product spaces which arise in analysis, most of them having inner products defined in terms of integrals.

1.3 Exercise   Show that the formula

$$(f, g) = \int_0^1 f(t)\overline{g(t)}\, dt$$

defines an inner product on the complex vector space $C[0, 1]$ of all continuous $\mathbb{C}$-valued functions on $[0, 1]$, with pointwise addition and scalar multiplication. $\square$

1.4 Exercise   Show that the formula

$$(A, B) = \text{trace}(B^* A)$$

defines an inner product on the space $\mathbb{C}^{m \times n}$ of $m \times n$ complex matrices, where $m, n \in \mathbb{N}$ and B^* denotes the conjugate transpose of B.   □

The conditions (ii) and (iii) in the definition of inner product are often summarized by the statement that $(\cdot, \cdot)$ is linear in the first argument. It follows from the definition that it is also *conjugate linear* in the second argument: this means that it satisfies (i) and (ii) of the following.

1.5 Theorem   For any x, y, z in an inner product space V and any $\lambda \in \mathbb{C}$,

(i) $(x, y + z) = (x, y) + (x, z)$;

(ii) $(x, \lambda y) = \bar{\lambda}(x, y)$;

(iii) $(x, 0) = 0 = (0, x)$;

(iv) if $(x, z) = (y, z)$ for all $z \in V$ then $x = y$.

Proof. (i) Using Definition 1.2(i) and (iii) we have

$$\begin{aligned}
(x, y + z) &= (y + z, x)^- \\
&= [(y, x) + (z, x)]^- \\
&= (y, x)^- + (z, x)^- \\
&= (x, y) + (x, z).
\end{aligned}$$

The proof of (ii) is similar. To prove (iii) put $\lambda = 0$ in (ii).

(iv) If $(x, z) = (y, z)$ then

$$\begin{aligned}
0 &= (x, z) + (-1)(y, z) \\
&= (x, z) + (-y, z) = (x - y, z).
\end{aligned}$$

If this holds for all $z \in V$ then in particular it holds when $z = x - y$; thus $(x - y, x - y) = 0$. By 1.2(iv) it follows that $x - y = 0$.   □

1.1   Inner product spaces as metric spaces

In the familiar case of $\mathbb{R}^3$ the magnitude $|u|$ of a vector u is equal to $(u, u)^{1/2}$, and the Euclidean distance between points with position vectors u, v is $|u - v|$. We copy this to introduce a natural metric in an inner product space.

1.6 Definition   The *norm* of a vector x in an inner product space is defined to be $(x, x)^{1/2}$. It is written $\|x\|$.   □

Thus, for $x = (x_1, \ldots, x_n) \in \mathbb{C}^n$ we have

$$\|x\| = (|x_1|^2 + \cdots + |x_n|^2)^{1/2},$$

while for $f \in C[0, 1]$, with the inner product described in Exercise 1.3,

$$\|f\| = \left\{ \int_0^1 |f(t)|^2 \, dt \right\}^{1/2}.$$

1.7 *Exercise* Let $x = (1/n)_{n=1}^{\infty} \in \ell^2$. Show that $\|x\| = \pi/\sqrt{6}$. What is $\|I_n\|$ where $I_n \in \mathbb{C}^{n \times n}$ is the identity matrix and the inner product of Exercise 1.4 is used?   □

1.8 *Theorem* For any x in an inner product space V and any $\lambda \in \mathbb{C}$

   (i) $\|x\| \geq 0$; $\|x\| = 0$ if and only if $x = 0$;

   (ii) $\|\lambda x\| = |\lambda| \|x\|$.

Proof. (ii)

$$\|\lambda x\| = (\lambda x, \lambda x)^{1/2} = \{\lambda \bar{\lambda}(x, x)\}^{1/2}$$
$$= |\lambda| \|x\|.$$ □

One knows that in $\mathbb{R}^3$ (x, y) is $\|x\| \|y\|$ times the cosine of an angle, from which it follows that $|(x, y)| \leq \|x\| \|y\|$. This relation continues to hold in a general inner product space.

1.9 *Theorem* For x, y in an inner product space V,

$$|(x, y)| \leq \|x\| \|y\|, \tag{1.3}$$

with equality if and only if x and y are linearly dependent.

(1.3) is known as the *Cauchy–Schwarz inequality*.

Proof. Suppose first that x and y are linearly dependent – say $x = \lambda y$ where $\lambda \in \mathbb{C}$. Then both sides of (1.3) equal $|\lambda| \|y\|^2$, and so (1.3) holds with equality.

Now suppose that x and y are linearly independent: we must show that (1.3) holds with strict inequality. For any $\lambda \in \mathbb{C}$, $x + \lambda y \neq 0$ and therefore

$$0 < (x + \lambda y, x + \lambda y)$$
$$= (x, x + \lambda y) + (\lambda y, x + \lambda y)$$
$$= (x, x) + (x, \lambda y) + (\lambda y, x) + (\lambda y, \lambda y)$$
$$= \|x\|^2 + \bar{\lambda}(x, y) + \lambda(x, y)^- + |\lambda|^2 \|y\|^2$$
$$= \|x\|^2 + 2 \operatorname{Re}\{\bar{\lambda}(x, y)\} + |\lambda|^2 \|y\|^2.$$

Pick a complex number u of unit modulus such that $\bar{u}(x, y) = |(x, y)|$. On putting $\lambda = tu$ we deduce that, for any $t \in \mathbb{R}$,

$$0 < \|x\|^2 + 2|(x, y)|t + \|y\|^2 t^2.$$

This can only happen if the real quadratic on the right hand side has negative discriminant: that is,

$$4|(x, y)|^2 - 4\|x\|^2\|y\|^2 < 0,$$

which yields the desired conclusion

$$|(x, y)| < \|x\|\|y\|.$$ $\square$

1.10 Exercise Prove that, for any $f \in C[0, 1]$,

$$\left| \int_0^1 f(t) \sin \pi t \, dt \right| \leqslant \frac{1}{\sqrt{2}} \left\{ \int_0^1 |f(t)|^2 \, dt \right\}^{1/2},$$

and describe the functions f for which equality holds. $\square$

The following relation is known as the *triangle inequality*.

1.11 Theorem For any x, y in an inner product space V,

$$\|x + y\| \leqslant \|x\| + \|y\|.$$

Proof. We have (compare the proof of Theorem 1.9)

$$\begin{aligned}
\|x + y\|^2 &= \|x\|^2 + 2 \operatorname{Re}(x, y) + \|y\|^2 \\
&\leqslant \|x\|^2 + 2|(x, y)| + \|y\|^2 \\
&\leqslant \|x\|^2 + 2\|x\|\|y\| + \|y\|^2 \\
&= (\|x\| + \|y\|)^2.
\end{aligned}$$ $\square$

1.12 Exercise By applying 1.11 to $\mathbb{C}^k, k = 1, 2, \ldots$, show that ℓ^2 is closed under addition. $\square$

1.13 Theorem (the parallelogram law) For vectors x, y in an inner product space,

$$\|x + y\|^2 + \|x - y\|^2 = 2\|x\|^2 + 2\|y\|^2.$$

Proof. We have

$$\|x + y\|^2 = \|x\|^2 + (x, y) + (y, x) + \|y\|^2,$$
$$\|x - y\|^2 = \|x\|^2 - (x, y) - (y, x) + \|y\|^2.$$

Add. $\square$

The reader is recommended to draw a diagram in order to see the reason for the name of this relation.

We have defined the norm in an inner product space in terms of the inner product; it has some significance that if we know how to calculate the norm of any vector then we can recover the inner product. This is because of the following result, called the *polarization identity*, the proof of which is a simple exercise.

1.14 **Theorem** For any x, y in an inner product space,

$$4(x, y) = \|x + y\|^2 - \|x - y\|^2 + i\|x + iy\|^2 - i\|x - iy\|^2. \qquad \square$$

Note that the polarization identity can be written

$$(x, y) = \frac{1}{4} \sum_{n=0}^{3} i^n \|x + i^n y\|^2.$$

1.15 **Exercise** Let H, K be inner product spaces and let $U: H \to K$ be a linear mapping such that $\|Ux\| = \|x\|$ for all $x \in H$. Prove that

$$(Ux, Uy) = (x, y) \qquad \text{for all } x, y \in H. \qquad \square$$

In the coming chapters we shall need a supply of examples on which to try out the concepts and results we shall meet. So far the only inner product spaces we know are $\mathbb{C}^n$, ℓ^2, $\mathbb{C}^{m \times n}$ and $C[0, 1]$. Calculations in the first three are often simple, but do not exhibit effectively the interaction of inner product properties with other mathematical structure. On the other hand, $C[0, 1]$ has the disadvantage that a general continuous function is a rather intangible entity. An ideal source of exercises is to be found in inner product spaces of rational functions, relatively concrete objects for which the reader will already have some intuition. Calculation in these spaces is often quite easy, particularly when use is made of basic complex analysis (principally Cauchy's integral formula, the residue theorem and some elementary facts about power series expansions). They illustrate the power of Hilbert space methods in complex analysis and lead on to the topic of the last quarter of the book, where we shall see that spaces of rational functions are of practical importance in a range of engineering applications.

1.16 **Examples** RL^2 denotes the space of rational functions which are analytic on the unit circle

$$\partial \mathbb{D} = \{z \in \mathbb{C} : |z| = 1\},$$

with the usual addition and scalar multiplication and with the inner product

$$(f, g) = \frac{1}{2\pi i} \int_{\partial \mathbb{D}} f(z) \overline{g(z)} \frac{dz}{z}, \qquad (1.4)$$

the integral being taken anti-clockwise round $\partial \mathbb{D}$.

RH^2 is the subspace of RL^2 consisting of those rational functions which are analytic on the closed unit disc clos $\mathbb{D}$, where

$$\mathbb{D} = \{z \in \mathbb{C} : |z| < 1\},$$

with inner product given by (1.4).

Thus a rational function (i.e. a ratio of two polynomials with complex coefficients) belongs to RL^2 provided it has no pole of modulus 1 and belongs to RH^2 if it has no pole of modulus less than or equal to 1. The spaces L^2 and H^2 we shall meet later: the prefix 'R' stands for 'rational'. This is a convention used in electrical engineering.

Let us check that (1.4) does define an inner product on RL^2. Axioms (i) to (iii) in Definition 1.2 are immediate; we must show that $(f, f) > 0$ when $f \neq 0$. On parametrizing $\partial\mathbb{D}$ by $z = e^{i\theta}$, $-\pi < \theta \leqslant \pi$, we can re-write (1.4) in the form

$$(f, g) = \frac{1}{2\pi} \int_{-\pi}^{\pi} f(e^{i\theta}) \overline{g(e^{i\theta})} \, d\theta,$$

so that, in particular

$$(f, f) = \frac{1}{2\pi} \int_{-\pi}^{\pi} |f(e^{i\theta})|^2 \, d\theta.$$

Since $f(e^{i\theta})$ is continuous on $[-\pi, \pi]$, the right hand side is positive unless $f = 0$.

To illustrate how Cauchy's integral formula simplifies calculations in RL^2 let us work out the inner product of the functions

$$f(z) = \frac{1}{z - \alpha}, \qquad g(z) = \frac{1}{z - \beta}$$

where $|\alpha| < 1$, $|\beta| < 1$.

$$(f, g) = \frac{1}{2\pi i} \int_{\partial\mathbb{D}} \frac{1}{z - \alpha} \cdot \frac{1}{\bar{z} - \bar{\beta}} \cdot \frac{dz}{z}.$$

Since $z\bar{z} = 1$ on $\partial\mathbb{D}$, this yields

$$(f, g) = \frac{1}{2\pi i} \int_{\partial\mathbb{D}} \frac{1}{z - \alpha} \cdot \frac{1}{1 - \bar{\beta}z} \, dz$$

$$= \frac{1}{2\pi i} \int_{\partial\mathbb{D}} \frac{h(z)}{z - \alpha} \, dz$$

where $h(z) = (1 - \bar{\beta}z)^{-1}$. Since $|\beta| < 1$, h is analytic on clos $\mathbb{D}$, and so Cauchy's integral formula applies to give

$$(f, g) = h(\alpha)$$

$$= \frac{1}{1 - \bar{\beta}\alpha}. \qquad \qquad \square$$

Another worthy inner product space is $W[a, b]$, described in Problem 1.2. This is a representative of a whole class of inner product spaces central to the theory of differential equations.

We turn next to the development of the properties of the metric $\|x - y\|$ in an inner product space. In fact it will pay us to begin with a more general type of space.

1.2   Problems

1.1.   For which $s \in \mathbb{C}$ does the sequence $(n^{-s})_{n=1}^{\infty}$ belong to ℓ^2?

1.2.   Let $a < b$ in $\mathbb{R}$. Show that the space $W[a, b]$ of continuously differentiable functions on $[a, b]$, with values in $\mathbb{C}$, is an inner product space with respect to pointwise addition and scalar multiplication, and inner product

$$(f, g)_W = \int_a^b f(t)\overline{g(t)} + f'(t)\overline{g'(t)} \, dt.$$

1.3.   A *trigonometric polynomial* is a function of the form

$$f(x) = \sum_{n=1}^k a_n e^{i\lambda_n x}$$

where $k \in \mathbb{N}$, $a_1, \ldots, a_k \in \mathbb{C}$ and $\lambda_1, \ldots, \lambda_k \in \mathbb{R}$. The space TP of trigonometric polynomials is a vector space over $\mathbb{C}$ with respect to pointwise addition and scalar multiplication. Prove that the formula

$$(f, g) = \lim_{T \to \infty} \frac{1}{2T} \int_{-T}^T f(x)\overline{g(x)} \, dx$$

defines an inner product on TP.

1.4.   Let V be an inner product space and let $x, y \in V$. Prove Pythagoras' theorem: if $(x, y) = 0$ then

$$\|x + y\|^2 = \|x\|^2 + \|y\|^2.$$

1.5.   Show that if x, y are vectors in an inner product space and $\|x + y\| = \|x\| + \|y\|$ then one of x, y is a scalar multiple of the other.

1.6.   Let $e_n(t) = e^{2\pi i n t}$, $n \in \mathbb{Z}$, $0 \leqslant t \leqslant 1$. Prove that $(e_n, e_m) = 0$ when $n \neq m$ with respect to the inner products of both $C[0, 1]$ (Exercise 1.3) and $W[0, 1]$ (Problem 1.2). Hence show that the values of $\|e_n - e_m\|$, $n \neq m$, with respect to the two inner products are $\sqrt{2}$ and $\sqrt{\{2 + 4\pi^2(n^2 + m^2)\}}$ respectively.

1.7.   Find an inner product on the space of complex polynomials such that the corresponding norm is given by

$$\|f\| = \left\{ \int_{-1}^1 |x| |f(x)|^2 + 3|f'(x)|^2 \, dx \right\}^{1/2}.$$

1.8.   Prove that, for any continuously differentiable function f on $[-\pi, \pi]$,

$$\left| \int_{-\pi}^{\pi} f(t) \cos t - f'(t) \sin t \, dt \right| \leqslant \sqrt{(2\pi)} \left\{ \int_{-\pi}^{\pi} |f(t)|^2 + |f'(t)|^2 \, dt \right\}^{1/2}.$$

1.9.   Prove that, for any polynomial f,

$$\left| \int_{-1}^1 |x|^3 f(x) + 6xf'(x) \, dx \right| \leqslant \frac{5}{\sqrt{3}} \left\{ \int_{-1}^1 |x| |f(x)|^2 + 3|f'(x)|^2 \, dx \right\}^{1/2}.$$

1.10. Find an analogue of the polarization identity in which i is replaced by a kth root of unity, $k \geqslant 3$.

1.11. Let k_α in RL^2 be defined by

$$k_\alpha(z) = (1 - \bar{\alpha}z)^{-1},$$

where $|\alpha| \neq 1$. Show that, for $f \in RH^2$,

$$(f, k_\alpha) = \begin{cases} f(\alpha) & \text{if } |\alpha| < 1 \\ 0 & \text{if } |\alpha| > 1. \end{cases}$$

1.12. Show that, for $f \in RH^2$ and $\alpha \in \mathbb{D}$,

$$|f(\alpha)| \leqslant \frac{1}{\sqrt{(1 - |\alpha|^2)}} \left\{ \frac{1}{2\pi} \int_{-\pi}^{\pi} |f(e^{i\theta})|^2 \, d\theta \right\}^{1/2}.$$

1.13. Let α, β be distinct points in $\mathbb{D}$ and let

$$g(z) = \frac{1}{(z - \alpha)(z - \beta)}.$$

Use the residue theorem to show that

$$\|g\| = \frac{(1 - |\alpha\beta|^2)^{1/2}}{(1 - |\alpha|^2)^{1/2}(1 - |\beta|^2)^{1/2}|1 - \bar{\alpha}\beta|}$$

in RL^2.

2

Normed spaces

Inner product spaces are not the only spaces of functions which arise naturally in analysis and admit a notion of distance. For example, in the space $C[a, b]$ of continuous functions on $[a, b]$ we could take the distance between functions f and g to be the maximum modulus of $f(x) - g(x)$ for $x \in [a, b]$. Convergence with respect to this metric is the same as uniform convergence, so it is certainly a natural notion; however, this metric determines a different topology from the integral inner product of Exercise 1.3, and moreover there is *no* inner product on $C[a, b]$ which gives rise to the maximum modulus metric (see Problems 2.3 and 2.4). Although we are concerned principally with inner product spaces, we shall be forced to make some use of other metrics. We shall therefore treat the elementary topological properties of function spaces with metrics in a broader context. We have seen in Theorems 1.8 and 1.11 that the norm in an inner product space satisfies the following conditions:

N1. $\|x\| > 0$ if $x \neq 0$;
N2. $\|\lambda x\| = |\lambda| \|x\|$ for all scalars λ and vectors x;
N3. $\|x + y\| \leqslant \|x\| + \|y\|$ for all vectors x, y.

2.1 Definition Let E be a real or complex vector space. A *norm* on E is a mapping $\| \cdot \| : E \to \mathbb{R}$ which satisfies N1, N2 and N3 above. A *normed space* is a pair $(E, \| \cdot \|)$ where E is a real or complex vector space and $\| \cdot \|$ is a norm on E. $\qquad\square$

Note that on putting $\lambda = 0$ in N2 we obtain $\|0\| = 0$ in any normed space.

2.2 Example The *supremum norm* $\| \cdot \|_\infty$ on the complex vector space $C(X)$ of all bounded continuous $\mathbb{C}$-valued functions on a topological space

X is defined by

$$\|f\|_\infty = \sup_{x \in X} |f(x)|.$$ ☐

It is elementary to check N1–N3 for this norm.

2.3 Theorem In any normed space $(E, \|\cdot\|)$ the function $d: E \times E \to \mathbb{R}$ defined by

$$d(x, y) = \|x - y\|$$

is a translation-invariant metric.

To say that d is translation-invariant means that translation of a pair of points by the same vector leaves their distance unchanged; in other words,

$$d(x + z, y + z) = d(x, y)$$

for all $x, y, z \in E$.

Proof. Translation-invariance is trivial. We have

$$d(x, x) = \|x - x\| = \|0\| = 0,$$

and, from N1, $d(x, y) > 0$ when $x \neq y$. By N2, $d(x, y) = d(y, x)$. By N3, for any vectors x, y, z,

$$\begin{aligned} d(x, z) = \|x - z\| &= \|x - y + y - z\| \\ &\leqslant \|x - y\| + \|y - z\| \\ &= d(x, y) + d(y, z). \end{aligned}$$ ☐

Thus every normed space (and so, in particular, every inner product space) is a metric space, so that we may make use of standard notions in the topology of metric spaces (see, for example, Binmore, 1981). This has proved to be an essential step in the study of function spaces: mathematicians discovered long ago that a purely algebraic theory of infinite-dimensional vector spaces was not adequate. Spaces which are provided with a topology, but not necessarily a metric, are also much used and studied, but they are not appropriate to an introductory course and we shall stick to normed spaces.

2.4 Exercise Prove that, for any normed space $(E, \|\cdot\|)$, the mapping $\|\cdot\|: E \to \mathbb{R}$ is continuous ('the norm is a continuous function'). ☐

It will be taken for granted that, in statements like the above, the topology on E being alluded to is the one defined by the metric $d(x, y) = \|x - y\|$.

The topological and algebraic structures of a normed space are related.

2.5 *Theorem* In a normed space E the algebraic operations are continuous: that is, addition and scalar multiplication are continuous as mappings

$$E \times E \to E,$$

$$k \times E \to E,$$

where k denotes the scalar field of E ($\mathbb{R}$ or $\mathbb{C}$) and $E \times E, k \times E$ have their respective product topologies.

Proof. Let us prove the continuity of scalar multiplication at the point $\langle \lambda, x \rangle$ of $k \times E$. For any $\mu \in k$, $y \in E$ we have

$$\|\lambda x - \mu y\| = \|\lambda x - \mu x + \mu x - \mu y\|$$
$$\leqslant \|\lambda x - \mu x\| + \|\mu x - \mu y\|$$
$$= |\lambda - \mu| \|x\| + |\mu| \|x - y\|. \tag{2.1}$$

Consider any $\varepsilon > 0$. Let

$$U_1 = \left\{ \mu \in k : |\mu - \lambda| < \min\left(1, \frac{\varepsilon}{2(1 + \|x\|)}\right) \right\},$$

$$U_2 = \left\{ y \in E : \|y - x\| < \frac{\varepsilon}{2(1 + |\lambda|)} \right\}.$$

Then $U_1 \times U_2$ is an open neighbourhood of $\langle \lambda, x \rangle$ in $k \times E$, and, for $\langle \mu, y \rangle \in U_1 \times U_2$ we have

$$|\mu| < |\lambda| + 1$$

and, in view of (2.1),

$$\|\lambda x - \mu y\| \leqslant \frac{\varepsilon}{2(1 + \|x\|)} \|x\| + |\mu| \frac{\varepsilon}{2(1 + |\lambda|)}$$
$$< \varepsilon. \qquad \square$$

The proof of the continuity of addition is similar, as is the proof of the following statement.

2.6 *Theorem* In an inner product space V the inner product is a continuous mapping from $V \times V$ with the product topology to $\mathbb{C}$. $\square$

Although it is straightforward enough to prove this directly, we observe that it can also be deduced from the polarization identity and the continuity of the algebraic operations and the norm.

2.1 Closed linear subspaces

In $\mathbb{C}^n$ every linear subspace is closed. In infinite-dimensional normed spaces it is not so.

2.7 Example   Let ℓ_0 denote the space of sequences $x = (x_n) \in \mathbb{C}^\mathbb{N}$ which have only finitely many terms different from zero. It is easy to check that ℓ_0 is a linear subspace of ℓ^2. It is not, however, a closed set in ℓ^2. For each $k \in \mathbb{N}$ the sequence

$$x^k = \left(1, \frac{1}{2}, \frac{1}{3}, \dots, \frac{1}{k}, 0, 0, \dots\right)$$

belongs to ℓ_0, and the sequence $(x^k)_{k=1}^\infty$ of vectors in ℓ^2 converges to

$$a = \left(1, \frac{1}{2}, \frac{1}{3}, \dots\right) = \left(\frac{1}{n}\right)_{n=1}^\infty$$

in ℓ^2 as $k \to \infty$, since

$$\|x^k - a\| = \left\|\left(0, 0, \dots, 0, \frac{1}{k+1}, \dots\right)\right\|$$
$$= \left\{\sum_{n=k+1}^\infty \frac{1}{n^2}\right\}^{1/2} \to 0.$$

The limit point a does not belong to ℓ_0, and so ℓ_0 is not closed.   □

2.8 Exercise   Show that

$$F = \{f \in C[0, 1] : f(0) = 0\}$$

is a subspace of $C[0, 1]$ which is not closed with respect to the inner product of Exercise 1.3.   □

From now on 'subspace' will always mean 'linear subspace' when it relates to a subset of a vector space. If you cannot do the exercise, turn to Problem 2.5.

2.9 Theorem   The closure of a subspace of a normed space is a subspace. *Proof.* Let F be a subspace of a normed space E and let clos F denote its closure. Let $x, y \in \text{clos } F$. There exist sequences $(x_n), (y_n)$ in F such that $x_n \to x$, $y_n \to y$. It follows from the continuity of addition that $x_n + y_n \to x + y$. Now $x_n + y_n \in F$ for each n, since F is a subspace, and hence $x + y \in \text{clos } F$. Thus clos F is closed under addition. Likewise it is closed under scalar multiplication, and so is a subspace of E.   □

2.10 Definition   Let $(E, \|\cdot\|)$ be a normed space and let $A \subseteq E$.

The *linear span of A*, lin A, is the intersection of all subspaces of E which contain A.

The *closed linear span of A*, clin A, is the intersection of all closed linear subspaces of E which contain A.   □

2.11 Exercise   Show that lin A is the unique subspace of E which contains A, and is contained in every subspace of E which contains A. Show likewise that clin A is the unique closed subspace of E which contains A and is contained in every closed subspace of E which contains A.   □

We can give a more concrete description of lin A. Let F denote the set of all (finite) linear combinations of elements of A: that is, if k denotes the scalar field of E,

$$F = \left\{ \sum_{n=1}^{m} \lambda_n a_n : m \in \mathbb{N}, \lambda_1, \ldots, \lambda_m \in k, a_1, \ldots, a_m \in A \right\}.$$

F is clearly a linear subspace of E which contains A, so that lin $A \subseteq F$. On the other hand, every linear subspace of E which contains A must also contain F, so that lin $A \supseteq F$. Hence lin $A = F$. In conjunction with the following result, this helps to form a picture of clin A, of sorts.

2.12 Theorem   For any set A in a normed space E, clin A is the closure of lin A.

Proof. By Theorem 2.9, clos lin A is a subspace of E. It is closed and contains A, and so, by the definition of clin A, clos lin $A \supseteq$ clin A.

On the other hand, clin A is a closed set in E which contains lin A. Hence it also contains clos lin A: clin $A \supseteq$ clos lin A.   □

Topological questions in finite-dimensional normed spaces are no harder than in the familiar space $\mathbb{R}^n$. Indeed, they are exactly the same.

2.13 Theorem   Any two norms on a finite-dimensional space E over $\mathbb{R}$ or $\mathbb{C}$ determine the same topology.

Proof. Let $e_1, \ldots, e_n$ be a basis of E. We can define a special ('Euclidean') norm ρ on E by the formula

$$\rho\left(\sum_1^n \lambda_j e_j \right) = \left\{ \sum_1^n |\lambda_j|^2 \right\}^{1/2}.$$

We shall show that if $\| \cdot \|$ is any norm on E then $\| \cdot \|$ and ρ determine the same topology. If $x = \sum_1^n \lambda_j e_j$ then

$$\|x\| = \left\| \sum_1^n \lambda_j e_j \right\| \leqslant \sum_1^n \| \lambda_j e_j \|$$

$$= \sum_1^n |\lambda_j| \| e_j \|$$

$$\leqslant \left\{ \sum_1^n |\lambda_j|^2 \right\}^{1/2} \left\{ \sum_1^n \| e_j \|^2 \right\}^{1/2}$$

$$= M \rho(x)$$

where

$$M = \left\{ \sum_1^n \|e_j\|^2 \right\}^{1/2} > 0.$$

It follows that, for any $\varepsilon > 0$, the open $\|\cdot\|$-ball of radius ε about $x_0 \in E$ contains the ρ-ball of radius ε/M about x_0. Hence every $\|\cdot\|$-open set in E is also ρ-open.

Define a function f from $\mathbb{R}^n$ or $\mathbb{C}^n$ to $\mathbb{R}$ by

$$f(\lambda_1, \ldots, \lambda_n) = \left\| \sum_1^n \lambda_j e_j \right\|.$$

This function is clearly continuous with respect to the natural Euclidean topology, and hence it attains its infimum $m \geq 0$ on the compact set

$$\left\{ (\lambda_1, \ldots, \lambda_n) : \sum_1^n |\lambda_j|^2 = 1 \right\}$$

(this depends on the Heine–Borel theorem and a property of continuous functions on compact sets: see Binmore [1981, pp. 71 and 74]). Since $e_1, \ldots, e_n$ are linearly independent m cannot be zero: hence $m > 0$. We have shown that $\|x\| \geq m > 0$ whenever $\rho(x) = 1$. It follows by homogeneity that

$$\|x\| \geq m\rho(x), \qquad \text{all } x \in E.$$

For any $\varepsilon > 0$, the ρ-ball of radius ε about $x_0 \in E$ contains the $\|\cdot\|$-ball of radius $m\varepsilon$ about x_0. Hence every ρ-open set is also $\|\cdot\|$-open. $\qquad\square$

2.14 Corollary Closed bounded sets in a finite-dimensional normed space are compact.

Proof. Let $(E, \|\cdot\|)$ be a finite-dimensional normed space and let K be a closed bounded set in E. In the notation of the foregoing proof, since

$$\rho(x) \leq m^{-1}\|x\|, \qquad \text{all } x \in E,$$

K is also bounded with respect to ρ. By the identity of topologies, K is also closed with respect to ρ. Since K is closed and bounded in the Euclidean space (E, ρ) it is compact, by the Heine–Borel theorem. $\qquad\square$

2.2   Problems

2.1. ℓ^∞ denotes the complex vector space of all bounded sequences $x = (x_n)_1^\infty$ of complex numbers, with componentwise addition and scalar multiplication. Verify that

$$\|x\|_\infty = \sup_{n \in \mathbb{N}} |x_n|$$

defines $\|\cdot\|_\infty$ as a norm on ℓ^∞.

2.2. ℓ^1 denotes the complex vector space of all absolutely summable sequences of complex numbers: that is, all sequences $x = (x_n)_1^\infty$, with $x_n \in \mathbb{C}$, such that

$$\sum_{n=1}^{\infty} |x_n| < \infty,$$

with componentwise addition and scalar multiplication. Show that

$$\|x\|_1 = \sum_{n=1}^{\infty} |x_n|$$

defines $\|\cdot\|_1$ as a norm on ℓ^1.

2.3. Prove that the supremum norm $\|\cdot\|_\infty$ of Example 2.2 and the inner product of Exercise 1.3 define different topologies on $C[0, 1]$ (find a sequence of functions (f_n) such that $\|f_n\|_\infty = 1$ for all n but $(f_n, f_n) \to 0$ as $n \to \infty$).

2.4. Prove that there is no inner product on $C[0, 1]$ such that $(f, f)^{1/2} = \|f\|_\infty$ for all f (show that the parallelogram law does not hold for $\|\cdot\|_\infty$).

2.5. Prove that, for $\alpha \in \mathbb{D}$,

$$\{f \in RH^2 : f(\alpha) = 0\}$$

is a closed linear subspace of RH^2 (use Problem 1.11).

2.6. Show that the subspace of polynomial functions is not closed in $C[0, 1]$ with respect to either the supremum norm or the inner product of Exercise 1.3.

2.7. c_0 denotes the subspace of ℓ^∞ comprising all sequences (x_n) which tend to zero as $n \to \infty$. Prove that c_0 is closed in ℓ^∞ with respect to $\|\cdot\|_\infty$.

2.8. The *open* and *closed unit balls* in a normed space $(E, \|\cdot\|)$ are defined to be respectively the sets

$$E_0 = \{x \in E : \|x\| < 1\}, \qquad E_1 = \{x \in E : \|x\| \leqslant 1\}.$$

Show that the latter is the closure of the former.

2.9. Show that c_0 (Problem 2.7) is the closed linear span in ℓ^∞ of the set $\{e_n : n \in \mathbb{N}\}$, where e_n is the sequence with nth term 1 and all other terms zero.

2.10. Guess the closed linear span in $(C[0, 1], \|\cdot\|_\infty)$ of the functions $\{1, x, x^2, \ldots\}$. The answer is given by *Weierstrass' approximation theorem* (Theorem 5.8 below).

2.11. Prove that in $W[0, 1]$ (Problem 1.2)

$$(f, \cosh) = f(1) \sinh 1$$

and deduce that

$$\{f \in W[0, 1] : f(1) = 0\}$$

is a closed subspace of $W[0, 1]$.

2.12. Give an example of an inner product space V and a dense subspace E of V which is of co-dimension 1 in V (this means that V/E is one dimensional, or in other words, that there exists $z \in V$ such that every element of V can be written in the form $e + \lambda z$ for some $e \in E$ and some scalar λ).

2.13. Prove that every finite-dimensional subspace of a normed space is closed.

2.14. Show that RH^2 is closed in RL^2.

2.15. Let the equation

$$z^3 + a_2 z^2 + a_1 z + a_0 = 0$$

(where $a_0, a_1, a_2 \in \mathbb{C}$) have distinct complex roots $\alpha_1, \alpha_2, \alpha_3$, each of modulus less than 1. Let

$$u_j = (1, \alpha_j, \alpha_j^2, \alpha_j^3, \ldots) \in \ell^2$$

for $j = 1, 2, 3$. Show that the linear span of $\{u_1, u_2, u_3\}$ is the solution space of the linear recurrence relation

$$x_{n+3} + a_2 x_{n+2} + a_1 x_{n+1} + a_0 x_n = 0,$$

$n = 1, 2, \ldots$.

3

Hilbert and Banach spaces

The inner product spaces $\mathbb{C}^n$ and ℓ^2 share a further convenient property: informally speaking, any sequence in either of these spaces which looks convergent *is* convergent. To see what this means, recall the inner product space ℓ_0 of Example 2.7. We saw there a sequence (x^k) converging in ℓ^2 but not in ℓ_0. If we tried to carry out all our analysis in ℓ_0 this phenomenon would definitely complicate matters: it would be like trying to do real analysis in $\mathbb{Q}$ instead of $\mathbb{R}$. Let us formulate the requirement of the existence of limits.

3.1 Definition   Let (M, d) be a metric space. A sequence (x^k) in M is a *Cauchy sequence* if, for every $\varepsilon > 0$, there exists an integer k_0 such that $k, l \geqslant k_0$ implies that $d(x^k, x^l) < \varepsilon$.

(M, d) is a *complete metric space* if every Cauchy sequence in M converges to a limit in M. □

Thus $\mathbb{R}$ is a complete metric space with respect to its natural metric. So also is $\mathbb{C}$, for if (z_k) is a Cauchy sequence in $\mathbb{C}$, then $(\operatorname{Re} z_k)$ and $(\operatorname{Im} z_k)$ are Cauchy sequences in $\mathbb{R}$. They thus have limits $x, y \in \mathbb{R}$, and we have $z_k \to x + iy$ in $\mathbb{C}$.

3.2 Theorem $\mathbb{C}^n$ (for $n \in \mathbb{N}$) and ℓ^2 are complete metric spaces.

We recall our convention that metric terminology refers to the metric determined by the norm according to Theorem 2.3 for an inner product space.

Proof. We prove the completeness of ℓ^2: the case of $\mathbb{C}^n$ is simpler.

Let (x^k) be a Cauchy sequence in ℓ^2. x^k is itself a sequence – say $x^k = (x_n^k)_{n=1}^\infty$.

We need to show that (x^k) converges in ℓ^2. The following three steps are standard in completeness proofs:

1. Find a candidate limit a;
2. Show that a belongs to the space in question;
3. Show that $x^k \to a$.

Step 1. Find a candidate limit a. We are looking for $a = (a_n)$ such that

$$x^1 = (x_1^1, x_2^1, x_3^1, \ldots, x_n^1, \ldots)$$
$$x^2 = (x_1^2, x_2^2, x_3^2, \ldots, x_n^2, \ldots)$$
$$\vdots$$
$$x^k = (x_1^k, x_2^k, x_3^k, \ldots, x_n^k, \ldots)$$

converges to

$$a = (a_1, a_2, a_3, \ldots, a_n, \ldots).$$

Consider a fixed 'column': that is, fix n. The sequence $(x_n^k)_{k=1}^{\infty}$ is a Cauchy sequence in $\mathbb{C}$. For consider any $\varepsilon > 0$. By hypothesis there exists $k_0 \in \mathbb{N}$ such that $k, l \geqslant k_0$ implies $\|x^k - x^l\| < \varepsilon$, and this certainly implies that $|x_n^k - x_n^l| < \varepsilon$, since it is clear that $|y_n| \leqslant \|y\|$ for any $y = (y_1, y_2, \ldots) \in \ell^2$ and any $n \in \mathbb{N}$. Since $\mathbb{C}$ is a complete metric space there exists $a_n \in \mathbb{C}$ such that $x_n^k \to a_n$ as $k \to \infty$.

Let $a = (a_1, a_2, \ldots) \in \mathbb{C}^{\mathbb{N}}$.

Step 2. The candidate limit belongs to ℓ^2. We show that $x^k - a \in \ell^2$ for some k. Since ℓ^2 is closed under subtraction it will follow that $a = x^k - (x^k - a) \in \ell^2$.

Let $\varepsilon > 0$. By hypothesis there exists $k_0 \in \mathbb{N}$ such that $k, k \geqslant k_0$ implies $\|x^k - x^l\| < \varepsilon$. Pick $N \in \mathbb{N}$: we have, for $k, l \geqslant k_0$,

$$\sum_{n=1}^{N} |x_n^k - x_n^l|^2 \leqslant \sum_{n=1}^{\infty} |x_n^k - x_n^l|^2$$
$$< \varepsilon^2.$$

Let $l \to \infty$ in the finite sum on the left hand side: we obtain

$$\sum_{n=1}^{N} |x_n^k - a_n|^2 \leqslant \varepsilon^2.$$

Since this holds for all $N \in \mathbb{N}$ we have, on letting $N \to \infty$,

$$\sum_{n=1}^{\infty} |x_n^k - a_n|^2 \leqslant \varepsilon^2,$$

that is, $x^k - a \in \ell^2$ and

$$\|x^k - a\| \leqslant \varepsilon.$$

Step 3. $x^k \to a$.

We have in fact shown in Step 2 that, for any $\varepsilon > 0$, there exists $k_0 \in \mathbb{N}$ such that $k \geqslant k_0$ implies $\|x^k - a\| \leqslant \varepsilon$. Hence $x^k \to a$.

Thus every Cauchy sequence in ℓ^2 converges to a limit in ℓ^2, and so ℓ^2 is complete. $\qquad\square$

3.3 Exercise Prove that ℓ^∞, the space of bounded sequences of complex numbers with the supremum norm (see Problem 2.1) is complete. $\qquad\square$

3.4 Definition A *Hilbert space* is an inner product space which is a complete metric space with respect to the metric induced by its inner product.

A *Banach space* is a normed space which is a complete metric space with respect to the metric induced by its norm. $\qquad\square$

Every Hilbert space is a Banach space, but there are plenty of Banach spaces which are not Hilbert spaces – for example, ℓ^∞. It is impossible to define an inner product on ℓ^∞ in such a way that $(x, x) = \|x\|_\infty^2$ for all $x \in \ell^\infty$: this follows from the failure of the parallelogram law in ℓ^∞.

In fact the parallelogram law characterizes Hilbert spaces in the class of Banach spaces. If a Banach space E does satisfy the parallelogram law then the formula

$$(x, y) = \frac{1}{4} \sum_{n=0}^{3} i^n \|x + i^n y\|^2$$

defines a genuine inner product on E. This is slightly tricky to prove.

Banach spaces are called after the Polish mathematician Stefan Banach, who played an important role in developing their theory and published an influential book on them, *Théorie des Opérations Linéaires* (1932). In recent years there have been remarkable discoveries about the structure of Banach spaces, but hitherto it has been the more special class of Hilbert spaces which has been central to applications of analysis.

3.1   The space $L^2(a, b)$

An example of an inner product space which is not a Hilbert space is $C[0, 1]$, the continuous functions on the unit interval with the inner product

$$(f, g) = \int_0^1 f(t)\overline{g(t)}\, dt, \tag{3.1}$$

described in Exercise 1.3. Problem 3.2 suggests how to construct a non-convergent Cauchy sequence in this space. Now we certainly want to use this inner product. It is the natural continuous version of the familiar inner product on $\mathbb{C}^n$, and it arises of itself in the context of integral equations and Fourier series. All that is wrong is the insistence on the continuity of f and *g*. Despite the incompleteness of $C[0, 1]$ Hilbert himself did always work with continuous functions, but since him mathematicians have followed F. Riesz in recognizing that the right setting for this inner product is the larger space $L^2(0, 1)$, which contains many discontinuous functions but is complete. Just as in passing from finite to infinite sequences we need to restrict ourselves to square-summable sequences to get a well-defined inner product, so in progressing further from an infinite sum to an integral we need to take square-integrable functions (that is, functions f for which $|f(t)|^2$ has a finite integral). To the distress of mathematicians of the time, the earliest attempts along these lines were unsuccessful. The set of Riemann integrable functions f on $(0, 1)$ such that

$$\int_0^1 |f(t)|^2 \, dt < \infty,$$

though much larger than $C[0, 1]$, is still incomplete with respect to the inner product (3.1). This was one of the reasons that analysts became dissatisfied with Riemann's definition of integral in the late nineteenth century. Many people worked at trying to develop a better integral, and the race was won by H. Lebesgue in 1902. His integral is much more complicated to define and develop than Riemann's, but as a tool it is easier to use as it has better properties, and in particular allows us to define a space of functions on which (3.1) is an inner product determining a complete metric. For the reader who is not familiar with the Lebesgue integral it is enough to accept that there is a definition of integral for which the space of square-integrable functions is complete with respect to (3.1). In later chapters we shall need some further properties of the Lebesgue integral, and these will be stated as required.

3.5 Example   The Hilbert space $L^2(a, b)$. Let $-\infty \leqslant a < b \leqslant \infty$. $L^2(a, b)$ is the space of Lebesgue measurable functions

$$f : (a, b) \to \mathbb{C}$$

which are square-integrable, in the sense that

$$\int_a^b |f(t)|^2 \, dt < \infty,$$

with pointwise operations and inner product

$$(f, g) = \int_a^b f(t)\overline{g(t)} \, dt. \tag{3.2}$$

The requirement that f be Lebesgue measurable is very mild, and for the purposes of this book can safely be ignored. It takes some ingenuity and one of the higher axioms of set theory to construct a function which is *not* Lebesgue measurable. Many engineers and scientists use the basics of L^2 theory throughout their careers without troubling themselves over this technicality: the functions they have to deal with are for the most part at worst mildly discontinuous.

Whatever integral one uses there is one subtlety to be discussed: which functions are to be regarded as equal? If f and g are functions in $L^2(a, b)$ which take the same values everywhere except at a finite number of points then

$$\|f - g\|^2 = \int_a^b |f(t) - g(t)|^2 \, dt = 0,$$

and hence we must regard f and g as equal in order to satisfy the axiom N1 for norms. More generally, we define a subset E of $\mathbb{R}$ to be a *null set* if, for every $\varepsilon > 0$, there exists a sequence $(a_n, b_n)_{n \in \mathbb{N}}$ of intervals such that $E \subseteq \bigcup_{n=1}^{\infty} (a_n, b_n)$ and $\sum_{n=1}^{\infty} b_n - a_n < \varepsilon$: in words, E is null if it can be covered by a sequence of open intervals of arbitrarily small total length. We then say that two functions f, g are equal *almost everywhere* if

$$\{t : f(t) \neq g(t)\} \quad \text{is a null set.}$$

In Lebesgue's theory, if f and g are equal almost everywhere then the integral of $|f - g|^2$ is zero, and so we must regard them as equal. Strictly speaking we should define the elements of $L^2(a, b)$ to be not functions but equivalence classes of functions equal almost everywhere. Conventionally, however, one speaks as if the elements were functions, with equality interpreted as equality almost everywhere. Once again, this is a technicality we can usually forget about, but notice that a function in L^2 need only be defined almost everywhere. Thus the formula $f(t) = |t|^{-1/4}$ defines an element of $L^2(-1, 1)$, even though $f(t)$ is undefined for $t = 0$.

It is an important fact that, for finite a and b, the space $C[a, b]$ of continuous $\mathbb{C}$-valued functions on $[a, b]$ is a dense subspace of $L^2(a, b)$ with respect to the L^2-norm. One can often prove that a statement holds for all $f \in L^2(a, b)$ by proving it for continuous f and using an approximation argument. A typical instance is the proof of Theorem 10.5 below.

For a proof that $L^2(a, b)$ is complete see any text on measure theory: for example de Barra (1981). Let us at least check that $L^2(a, b)$ is an inner product space. By the inequality of the means we have

$$2|f(t)\overline{g(t)}| \leqslant |f(t)|^2 + |g(t)|^2, \qquad a \leqslant t \leqslant b,$$

and so on integrating we deduce that, if $f, g \in L^2(a, b)$,

$$\int_a^b |f(t)\overline{g(t)}| \, dt < \infty \tag{3.3}$$

and hence (3.2) does define (f, g) as a complex number (its real and imaginary parts are both finite in view of (3.3)). And on integrating the inequality

$$|f(t) + g(t)|^2 \leqslant |f(t)|^2 + 2|f(t)g(t)| + |g(t)|^2$$

we can infer from (3.3) that $L^2(a, b)$ is closed under addition. It is obviously closed under scalar multiplication.   □

3.6 Exercise   For which real α does the appropriate restriction of the function $f(t) = t^\alpha$ belong to (i) $L^2(0, 1)$, (ii) $L^2(1, \infty)$, (iii) $L^2(0, \infty)$?   □

We come to a geometric property of Hilbert space which illustrates the claim that spatial intuition can be a tool in the understanding of functions.

3.2   The closest point property

If X is a point above a flat surface A then there is a unique point Y on A which is closer to X than any other point of A. Our years of familiarity with physical space make this seem an unexciting assertion, but when transferred to Hilbert space it yields a principle with far-reaching implications. In particular it underlies the whole of duality theory, through the proof of the Riesz–Fréchet theorem (6.8 below).

3.7 Definition   A subset A of a real or complex vector space is *convex* if, for all $a, b \in A$ and all λ such that $0 < \lambda < 1$, the point $\lambda a + (1 - \lambda)b$ belongs to A.   □

The set of points $\{\lambda a + (1 - \lambda)b : 0 < \lambda < 1\}$ should be thought of as the line segment joining a and b. A suitable mental picture to accompany the definition would be as shown in Diagram 3.1.

3.8 Theorem (The closest point property)   Let A be a non-empty closed convex set in a Hilbert space H. For any $x \in H$ there is a unique point of A which is closer to x than any other point of A.

That is, there is a unique point $y \in A$ such that

$$\|x - y\| = \inf_{a \in A} \|x - a\|.$$

Diagram 3.1

convex   not convex

Proof. Let

$$M = \inf_{a \in A} \|x - a\|.$$

Since $A \neq \varnothing$, $M < \infty$ and hence, for each $n \in \mathbb{N}$, there exists $y_n \in A$ such that

$$\|x - y_n\|^2 < M^2 + \frac{1}{n}. \tag{3.4}$$

We shall prove that (y_n) is a Cauchy sequence by applying the parallelogram law to the vectors $x - y_n$, $x - y_m$. We have, for any $m, n \in \mathbb{N}$,

$$\|x - y_n - (x - y_m)\|^2 + \|x - y_n + x - y_m\|^2$$
$$= 2\|x - y_n\|^2 + 2\|x - y_m\|^2$$
$$< 4M^2 + 2\left(\frac{1}{n} + \frac{1}{m}\right)$$

which can be re-arranged to give

$$\|y_n - y_m\|^2 < 4M^2 + 2\left(\frac{1}{n} + \frac{1}{m}\right) - 4\left\|x - \frac{y_n + y_m}{2}\right\|^2.$$

Since A is convex and $y_n, y_m \in A$, $(y_n + y_m)/2 \in A$, and therefore

$$\left\|x - \frac{y_n + y_m}{2}\right\|^2 \geq M^2.$$

Hence

$$\|y_n - y_m\|^2 < 4M^2 + 2\left(\frac{1}{n} + \frac{1}{m}\right) - 4M^2$$
$$= 2\left(\frac{1}{n} + \frac{1}{m}\right).$$

It follows that (y_n) is a Cauchy sequence, and so converges to a limit $y \in H$. Since A is closed, $y \in A$ and therefore

$$\|x - y\| \geq M.$$

On letting $n \to \infty$ in (3.4) we obtain

$$\|x - y\| \leq M,$$

so that $\|x - y\| = M$.

We have shown that there is a closest point to x in A: we must also prove it is unique. Suppose that $w \in A$ and $\|x - w\| = M$. Then $(y + w)/2 \in A$, so that

$$\left\|x - \frac{y + w}{2}\right\| \geq M.$$

On applying the parallelogram law to $x - w$, $x - y$ we obtain

$$\|y - w\|^2 = 2\|x - w\|^2 + 2\|x - y\|^2 - 4\left\|x - \frac{y + w}{2}\right\|^2$$

$$\leqslant 2M^2 + 2M^2 - 4M^2 = 0.$$

Thus $y = w$, and uniqueness is proved.   □

In a Banach space there may be infinitely many closest points (Problem 3.9) or none at all (Problems 6.6 and 6.7).

3.3   Problems

3.1.   Prove that a closed subspace of a complete metric space is itself complete with respect to the induced metric. Deduce that $(c_0, \| \cdot \|_\infty)$ is a Banach space (see Problem 2.7).

3.2. Prove that $C[0, 1]$ is not a Hilbert space with respect to the inner product defined in Exercise 1.3.

Hint: show that (f_n) is a non-convergent Cauchy sequence in $C[0, 1]$, where f_n has the graph shown in Diagram 3.2.

3.3. Show that the normed space $(C(X), \| \cdot \|_\infty)$ of Exercise 2.2 is a Banach space.

3.4. Show that $W[a, b]$ (Problem 1.2) is not a Hilbert space (use the indefinite integrals of the functions in Problem 3.2).

3.5. Show that RH^2 is not a Hilbert space.

3.6. For which real α does the function

$$f_\alpha(t) = t^\alpha e^{-t}, \qquad t > 0$$

belong to $L^2(0, \infty)$? What is $\| f_\alpha \|$, when defined?

3.7. Prove that the open and closed unit balls in a normed space are convex (Problem 2.8).

3.8. Prove that the closure of a convex set in a normed space is convex.

Diagram 3.2

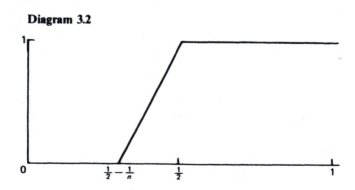

3.9. Let E be the Banach space $\mathbb{R}^2$ with norm

$$\|(x_1, x_2)\| = \max\{|x_1|, |x_2|\}.$$

Show that E does not have the closest point property by finding infinitely many points in the closed convex set

$$A = \{(x_1, x_2) : x_1 \geqslant 1\}$$

which are at minimal distance from the origin.

3.10. Let A be a non-empty closed convex set in a Hilbert space. Show that A contains a unique vector a of smallest norm and that $\mathrm{Re}(a, a - x) \leqslant 0$ for all $x \in A$.

3.11. Can you find a Banach space E, a closed convex set $A \subset E$ and a point $x \in E$ such that A contains no closest point to x: that is, so that

$$\inf_{y \in A} \|x - y\|_E$$

is not attained? One example is outlined in Problems 6.6 and 6.7.

4

Orthogonal expansions

The use of vectors to study the geometry of $\mathbb{R}^3$ enables us to represent a point by a single symbol r instead of seven characters (x, y, z). In this way we can give clearer and more elegant proofs of general theorems, but when it comes to applications we need a numerical representation of vectors, and so we revert to co-ordinates. Evidently there is a place for both the abstract notion of a vector and for co-ordinates. The same is true in Hilbert spaces of functions. In $\mathbb{R}^3$ one traditionally denotes by i, j, k mutually orthogonal unit vectors and writes

$$r = xi + yj + zk$$

where

$$x = r \cdot i, \qquad y = r \cdot j, \qquad z = r \cdot k.$$

One then has

$$|r|^2 = (r \cdot i)^2 + (r \cdot j)^2 + (r \cdot k)^2.$$

Let us copy this procedure.

4.1 Definition   Vectors x, y in an inner product space V are *orthogonal* (written $x \perp y$) if $(x, y) = 0$.

A family $(e_\alpha)_{\alpha \in A}$ in $V \backslash \{0\}$ is called an *orthogonal system* if $e_\alpha \perp e_\beta$ when $\alpha \neq \beta$. If, further, $\|e_\alpha\| = 1$ for each $\alpha \in A$, then the family $(e_\alpha)_{\alpha \in A}$ is called an *orthonormal system*.

An orthonormal system $(e_\alpha)_{\alpha \in A}$ is called an *orthonormal sequence* if it can be indexed by $\mathbb{N}$. $\qquad \square$

Note that a system indexed by $\mathbb{Z}$ can be 'renumbered' so that it is indexed by $\mathbb{N}$ instead: given $(e_n)_{n \in \mathbb{Z}}$ we can write $f_1 = e_0, f_2 = e_1, f_3 = e_{-1}, f_4 = e_2$, etc. Then $\{e_n : n \in \mathbb{Z}\} = \{f_n : n \in \mathbb{N}\}$. We may therefore take the typical orthonormal sequence to be indexed by $\mathbb{N}$ without loss of generality.

4.2 Examples In $\mathbb{C}^n$ the standard basis vectors constitute an orthonormal system; so does any subset of them.

In ℓ^2, $(e_n)_{n \in \mathbb{N}}$ is an orthonormal sequence, where e_n is the sequence in ℓ^2 having 1 in the nth place and zeros everywhere else.

In $L^2(-\pi, \pi)$ an orthonormal sequence is $(e_n)_{n \in \mathbb{Z}}$ where

$$e_n(t) = (2\pi)^{-1/2} e^{int},$$

for

$$(e_n, e_m) = \frac{1}{2\pi} \int_{-\pi}^{\pi} e^{i(n-m)t} \, dt = \begin{cases} 1 & \text{if } n = m \\ 0 & \text{otherwise.} \end{cases}$$

An alternative orthonormal sequence in $L^2(-\pi, \pi)$ is

$$(2\pi)^{-1/2}, \quad \pi^{-1/2} \cos t, \quad \pi^{-1/2} \sin t, \quad \pi^{-1/2} \cos 2t, \quad \dots \qquad \square$$

The latter examples explain the following terminology.

4.3 Definition If (e_n) is an orthonormal sequence in a Hilbert space H then, for any $x \in H$, (x, e_n) is the nth *Fourier coefficient of* x with respect to (e_n). The *Fourier series of* x with respect to the sequence (e_n) is the series $\sum_n (x, e_n) e_n$. $\qquad \square$

So far this is only a formal sum: we address now the problem of giving it a meaning. To put it another way, to what extent can orthogonal systems be made to play the role of co-ordinate systems? As a preliminary we note a generalization of Pythagoras' theorem.

4.4 Theorem If $x_1, \dots, x_n$ is an orthogonal system in an inner product space then

$$\left\| \sum_{j=1}^{n} x_j \right\|^2 = \sum_{j=1}^{n} \|x_j\|^2.$$

Proof. Write the left hand side as an inner product and expand. $\qquad \square$

The basic properties of orthogonal expansions are mainly derived from the following identity.

4.5 Lemma Let $e_1, \dots, e_n$ be an orthonormal system in an inner product space H, let $\lambda_1, \dots, \lambda_n \in \mathbb{C}$ and let $x \in H$. Then

$$\left\| x - \sum_{j=1}^{n} \lambda_j e_j \right\|^2 = \|x\|^2 + \sum_{j=1}^{n} |\lambda_j - c_j|^2 - \sum_{j=1}^{n} |c_j|^2$$

where $c_j = (x, e_j)$.

Proof. By Theorem 4.4, $(\sum \lambda_j e_j, \sum \lambda_j e_j) = \sum \lambda_j \bar{\lambda}_j$, and so

$$
\begin{aligned}
\left\| x - \sum \lambda_j e_j \right\|^2 &= (x - \sum \lambda_j e_j, x - \sum \lambda_j e_j) \\
&= (x, x) - \sum \lambda_j (e_j, x) - \sum \bar{\lambda}_j (x, e_j) + \sum \lambda_j \bar{\lambda}_j \\
&= \|x\|^2 - \sum \lambda_j \bar{c}_j - \sum \bar{\lambda}_j c_j + \sum \lambda_j \bar{\lambda}_j \\
&= \|x\|^2 + \sum (\lambda_j \bar{\lambda}_j - \lambda_j \bar{c}_j - \bar{\lambda}_j c_j + c_j \bar{c}_j) - \sum c_j \bar{c}_j \\
&= \|x\|^2 + \sum (\lambda_j - c_j)(\bar{\lambda}_j - \bar{c}_j) - \sum c_j \bar{c}_j \\
&= \|x\|^2 + \sum |\lambda_j - c_j|^2 - \sum |c_j|^2.
\end{aligned}
$$

Suppose now that x and the e_j are fixed and the λ_j are allowed to vary. Then $\sum \lambda_j e_j$ will trace out $\mathrm{lin}\{e_1, \ldots, e_n\}$. Since the c_j are fixed, it is clear from Lemma 4.5 that the smallest possible value of $\|x - \sum \lambda_j e_j\|$ occurs when $\lambda_j = c_j$, $1 \leqslant j \leqslant n$. We deduce the following statement.

4.6 Theorem Let $e_1, \ldots, e_n$ be an orthonormal system in an inner product space H and let $x \in H$. The closest point y of $\mathrm{lin}\{e_1, \ldots, e_n\}$ to x is

$$
y = \sum_{j=1}^{n} (x, e_j) e_j,
$$

and the distance $d = \|x - y\|$ is given by

$$
d^2 = \|x\|^2 - \sum_{j=1}^{n} |(x, e_j)|^2. \qquad \square
$$

4.7 Corollary If $x \in \mathrm{lin}\{e_1, \ldots, e_n\}$ then

$$
x = \sum_{j=1}^{n} (x, e_j) e_j. \qquad \square
$$

The reader should attempt to visualize Theorem 4.6 in terms of some such picture as Diagram 4.1.

Diagram 4.1

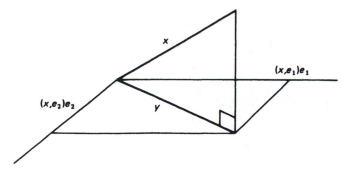

4.8 Exercise   In $L^2(0, 1)$ let $e_0(t) = 1$, $e_1(t) = \sqrt{3}(2t - 1)$ for all $t \in (0, 1)$. Show that e_0, e_1 is an orthonormal system in $L^2(0, 1)$, and show that the polynomial y of degree 1 which is closest with respect to the norm of $L^2(0, 1)$ to the function $x(t) = t^2$ is given by $y(t) = t - 1/6$. What is $\|x - y\|$?   □

4.1   Bessel's inequality

4.9   Theorem (Bessel's inequality)   If $(e_n)_{n \in \mathbb{N}}$ is an orthonormal sequence in an inner product space H then, for any $x \in H$,

$$\sum_{n=1}^{\infty} |(x, e_n)|^2 \leqslant \|x\|^2.$$

Proof. For $N \in \mathbb{N}$ let $y_N = \sum_{n=1}^{N} (x, e_n)e_n$. By Theorem 4.6,

$$\|x - y_N\|^2 = \|x\|^2 - \sum_{n=1}^{N} |(x, e_n)|^2,$$

and hence

$$\sum_{n=1}^{N} |(x, e_n)|^2 = \|x\|^2 - \|x - y_N\|^2$$
$$\leqslant \|x\|^2.$$

Let $N \to \infty$ in the last inequality.   □

We are hoping that the formal Fourier series $\sum (x, e_n)e_n$ will actually turn out to represent the vector x, and so we must say what we mean by the sum of an infinite series in a normed space. We take the most obvious generalization of the familiar notion for scalar series.

4.10 Definition   Let $(E, \|\cdot\|)$ be a normed space and let $x_n \in E$ for $n \in \mathbb{N}$. We say that $\sum_{n=1}^{\infty} x_n$ converges and has sum x, and write $\sum_{n=1}^{\infty} x_n = x$, if

$$\sum_{n=1}^{k} x_n \to x \qquad \text{as } k \to \infty.$$ □

In general it is difficult to ascertain whether a series in a Banach space converges, but in the case of series of orthogonal vectors in a Hilbert space there is a neat characterization.

4.11 Theorem   Let $(e_n)_{n \in \mathbb{N}}$ be an orthonormal sequence in a Hilbert space H and let $\lambda_n \in \mathbb{C}$, $n \in \mathbb{N}$. Then $\sum_{n=1}^{\infty} \lambda_n e_n$ converges in H if and only if

$$\sum_{n=1}^{\infty} |\lambda_n|^2 < \infty.$$

Proof. ($\Rightarrow$) Suppose $\sum_{n=1}^{\infty} \lambda_n e_n = x$. For $m \geqslant k$ in $\mathbb{N}$,

$$\left(\sum_{n=1}^{m} \lambda_n e_n, e_k \right) = \sum_{n=1}^{m} \lambda_n(e_n, e_k) = \lambda_k.$$

On letting $m \to \infty$ and using the continuity of the inner product we obtain

$$(x, e_k) = \lim_{m \to \infty} \lambda_k = \lambda_k.$$

By Bessel's inequality,

$$\sum_{n=1}^{\infty} |\lambda_n|^2 = \sum_{n=1}^{\infty} |(x, e_n)|^2 \leqslant \|x\|^2 < \infty.$$

($\Leftarrow$) Suppose $\sum_{n=1}^{\infty} |\lambda_n|^2 < \infty$ and write

$$x_m = \sum_{n=1}^{m} \lambda_n e_n.$$

By Pythagoras' theorem, for $m, p \in \mathbb{N}$,

$$\|x_{m+p} - x_m\|^2 = \left\| \sum_{n=m+1}^{m+p} \lambda_n e_n \right\|^2$$

$$= \sum_{n=m+1}^{m+p} |\lambda_n|^2$$

$$\to 0 \qquad \text{as } m \to \infty.$$

Thus (x_m) is a Cauchy sequence in H, and it therefore converges in H.   □

4.2   Pointwise and L^2 convergence

4.12   *Caution*   Suppose (f_n) is a sequence of functions in $L^2(0, 1)$. We have given a meaning to the statement

$$\sum_{n=1}^{\infty} f_n = f. \tag{4.1}$$

There is another interpretation of this statement which seems plausible, to wit,

$$\sum_{n=1}^{\infty} f_n(t) = f(t) \qquad \text{for all } t \in (0, 1).$$

In this case one says that the series (4.1) *converges pointwise*. (There is a commonly used modification in which one discounts 'set of measure zero'.) Be warned that the two notions of convergence are not the same. Here is a standard example of a sequence (g_n) which tends to zero with respect to the metric of $L^2(0, 1)$ but does not tend to zero pointwise. For a fixed $m = 0, 1, 2, \ldots$, divide $(0, 1)$ into 2^m equal intervals. Terms numbered 2^m up to $2^{m+1} - 1$ of the sequence (g_n) consist of the characteristic functions of each of these intervals in turn (the characteristic function of a set is the function which takes the value 1 on the set and is zero outside it; for the present purpose, the values at the end points of the intervals do not matter). Each of these terms has norm $2^{-m/2}$, so that $g_n \to 0$ in $L^2(0, 1)$, but for any $t \in (0, 1)$ not of the form $r/2^m$ with $r, m \in \mathbb{N}$, we have $g_n(t) = 1$ infinitely often.

A complementary example is obtained by taking h_n to be $\sqrt{n}$ times the characteristic function of the interval $(0, 1/n)$. Then $\|h_n\| = 1$ for each n, but $h_n(t) \to 0$ as $n \to \infty$ for each $t \in (0, 1)$.   □

A thoroughgoing study of these notions is part of measure theory; for the moment we only wish to register that they must be treated with care.

4.3   Complete orthonormal sequences

Let us resume the problem of giving a 'co-ordinate expression' for an element of a Hilbert space H. Given an orthonormal sequence (e_n) and a vector $x \in H$ we would like to write

$$x = \sum_{n=1}^{\infty} (x, e_n)e_n.$$

Bessel's inequality and Theorem 4.11 between them ensure that the right hand side makes sense: it is a convergent series in H. But without further assumptions on the (e_n) we cannot be sure that its limit is x. Consider, for example, the 'standard orthonormal sequence' (e_n) in ℓ^2, where e_n has nth term 1 and all other terms zero. Write $f_n = e_{n+1}$. Then $(f_n)_{n \in \mathbb{N}}$ is an orthonormal sequence in ℓ^2, and for $x = (x_n) \in \ell^2$ we have

$$\sum_{n=1}^{\infty} (x, f_n)f_n = \sum_{n=2}^{\infty} (x, e_n)e_n$$
$$= (0, x_2, x_3, \ldots).$$

In the general case, let us form the 'error'

$$y = x - \sum_{n=1}^{\infty} (x, e_n)e_n.$$

For each $j \in \mathbb{N}$ we have

$$(y, e_j) = (x, e_j) - \sum_{n=1}^{\infty} (x, e_n)(e_n, e_j)$$
$$= 0.$$

If we know that the only vector orthogonal to every e_n is the zero vector then we can infer that $y = 0$, and the desired representation is valid.

4.13 Definition   An orthonormal sequence (e_n) in a Hilbert space H is *complete* if the only member of H which is orthogonal to every e_n is the zero vector.   □

4.14 Theorem   Let $(e_n)_{n \in \mathbb{N}}$ be a complete orthonormal sequence in a Hilbert space H. For any $x \in H$,

$$x = \sum_{n=1}^{\infty} (x, e_n)e_n$$

and

$$\|x\|^2 = \sum_{n=1}^{\infty} |(x, e_n)|^2.$$

Proof. The first statement was proved above. By Pythagoras' theorem, for any $N \in \mathbb{N}$,

$$\left\| \sum_{n=1}^{N} (x, e_n) e_n \right\|^2 = \sum_{n=1}^{N} |(x, e_n)|^2.$$

On letting $N \to \infty$ and using the continuity of the norm we deduce the second statement. $\square$

A complete orthonormal sequence in H is also called an *orthonormal basis* of H.

This theorem presents a satisfactory answer of a general kind to the question raised earlier. In order to use it in a concrete instance, though, we need to know that a particular orthonormal sequence is complete. Showing this often requires some hard analysis.

The following characterization gives us a way of visualizing the completeness of an orthonormal sequence in topological terms.

4.15   Theorem   Let $(e_n)_{n \in \mathbb{N}}$ be an orthonormal sequence in a Hilbert space H. The following are equivalent:

   (1) $(e_n)_{n \in \mathbb{N}}$ is complete;

   (2) $\operatorname{clin}\{e_n : n \in \mathbb{N}\} = H$;

   (3) $\|x\|^2 = \sum_{n=1}^{\infty} |(x, e_n)|^2 \qquad$ for all $x \in H$.

Proof. (1) $\Rightarrow$ (2) and (1) $\Rightarrow$ (3) are contained in Theorem 4.14, since it is clear from the definition of convergence of a series that

$$\sum_{n=1}^{\infty} (x, e_n) e_n \in \operatorname{clin}\{e_n : n \in \mathbb{N}\}$$

for any $x \in H$.

(3) $\Rightarrow$ (1) Suppose (e_n) is not complete: then there is a non-zero $x \in H$ such that $x \perp e_n$ for all n. Then $\|x\| \neq 0$ but

$$\sum_{n=1}^{\infty} |(x, e_n)|^2 = 0,$$

so that (3) is violated.

(2) $\Rightarrow$ (1) Suppose (2) holds, and consider any $x \in H$ such that $x \perp e_n$ for all n. Let

$$E = \{y \in H : (x, y) = 0\}.$$

E is a subspace of H, and by the continuity of the inner product it is closed. Since it contains each e_n, it contains also $\mathrm{clin}\{e_n : n \in \mathbb{N}\} = H$. In particular, $x \in E$: that is, $(x, x) = 0$, and hence $x = 0$. Thus (e_n) is complete. $\qquad\square$

Statements analogous to the above hold for complete orthonormal *systems*, not just sequences. For many purposes it is sufficient to restrict attention to the conceptually simpler notion of sequence; most, but not all, of the naturally arising Hilbert spaces do possess complete orthonormal sequences.

4.16 Definition   A Hilbert space is *separable* if it contains a complete orthonormal sequence (indexed by $\mathbb{N}$ or finite). $\qquad\square$

In a sense we have already encountered all separable Hilbert spaces. To make this statement precise we need to formulate the notion of 'isomorphism' appropriate to Hilbert space.

4.17 Definition   A mapping $U : H \to K$, where H, K are Hilbert spaces is a *unitary operator* if it is linear and bijective and preserves inner products: that is, satisfies
$$(Ux, Uy) = (x, y)$$
for all $x, y \in H$.

Hilbert spaces H, K are *isomorphic* if there is a unitary operator from H to K. $\qquad\square$

4.18 Theorem   Let $U : H \to K$ be a linear and surjective mapping between Hilbert spaces. Then U is unitary if and only if $\|Ux\| = \|x\|$ for all $x \in H$.
Proof. The forward implication is immediate. The converse follows from Exercise 1.15: it depends on the polarization identity. $\qquad\square$

4.19 Theorem   Let H be a separable Hilbert space. H is isomorphic either to $\mathbb{C}^n$ for some $n \in \mathbb{N}$ or to ℓ^2.
Proof. Suppose H contains a finite complete orthonormal sequence $e_1, \ldots, e_n$. For any $x \in H$, the vector
$$x - \sum_{j=1}^{n} (x, e_j) e_j$$
is orthogonal to each e_j, hence is zero. Thus $e_1, \ldots, e_n$ is a basis of H. Define $U : H \to \mathbb{C}^n$ by
$$U(\xi_1 e_1 + \cdots + \xi_n e_n) = (\xi_1, \ldots, \xi_n).$$
Then U is linear and bijective. From the fact that $\|x\|^2 = \sum_{j=1}^{n} |(x, e_j)|^2$ we infer that $\|Ux\| = \|x\|$ for all $x \in H$. Thus U is a unitary operator, and so H is isomorphic to $\mathbb{C}^n$.

Now consider the alternative possibility: H contains a complete orthonormal sequence $(e_n)_{n \in \mathbb{N}}$. Define $U: H \to \ell^2$ by $Ux = (\xi_n)_{n \in \mathbb{N}}$ where $\xi_n = (x, e_n)$. Theorem 4.15 shows that Ux does belong to ℓ^2 and that $\|Ux\| = \|x\|$ for all $x \in H$. U is clearly linear. If $(\xi_n) \in \ell^2$ then, by Theorem 4.11, the series $\sum \xi_n e_n$ converges to a point $x \in H$, and we have $Ux = (\xi_n)$. Thus U is surjective. Hence U is a unitary operator and so H is isomorphic to ℓ^2. $\qquad\square$

In applications it is not just the isomorphism type which is of interest: we may be dealing with Hilbert spaces of, say, analytic or differentiable functions, and it is the interplay of the Hilbert space structure and the properties of individual functions which is fruitful. However, it is valuable to know that, as far as Hilbert space properties are concerned, we can just think of ℓ^2 or $\mathbb{C}^n$ when we wish.

4.4 Orthogonal complements

One can use orthogonality to split a Hilbert space up into a 'sum' of subspaces.

4.20 Definition The *orthogonal complement* of a subset E of a Hilbert space H is the set

$$\{x \in H : (x, y) = 0 \text{ for all } y \in E\}.$$

It is denoted by $H \ominus E$ or $E^\perp$, usually read 'E perp'. $\qquad\square$

4.21 Exercise Let

$$E = \{f \in L^2(0, 1) : f(t) = 0 \text{ for } 0 < t < \tfrac{1}{2}\}.$$

Describe the orthogonal complement of E in $L^2(0, 1)$. $\qquad\square$

4.22 Theorem For any set E in a Hilbert space H, $E^\perp$ is a closed linear subspace of H. $\qquad\square$

The proof is a routine exercise.

The 'perp' of a linear subspace, though defined in terms of the inner product, can be characterized in terms of the norm. The relevant observation in $\mathbb{R}^3$ is that a line segment OX is orthogonal to a plane M through O when O is the closest point to X in M.

4.23 Lemma Let M be a linear subspace of an inner product space H and let $x \in H$. Then $x \in M^\perp$ if and only if

$$\|x - y\| \geqslant \|x\| \qquad \text{for all } y \in M. \tag{4.2}$$

An appropriate picture is shown in Diagram 4.2.

Proof. ($\Rightarrow$) If $x \in M^\perp$ then, for any $y \in M$, x and y are orthogonal so that Pythagoras' theorem yields

$$\|x - y\|^2 = \|x\|^2 + \|y\|^2 \geqslant \|x\|^2.$$

($\Leftarrow$) Suppose that (4.2) holds. Pick any $y \in M$ and $\lambda \in \mathbb{C}$. Then $\lambda y \in M$ and so

$$\|x - \lambda y\|^2 \geqslant \|x\|^2.$$

Write the left hand side as an inner product, expand and cancel $\|x\|^2$ from each side to give

$$-2 \operatorname{Re} \bar{\lambda}(x, y) + |\lambda|^2 \|y\|^2 \geqslant 0.$$

Since this holds for all $\lambda \in \mathbb{C}$, in particular it holds for $\lambda = tz$ where $t > 0$ and z is a complex number of unit modulus chosen so that $\bar{z}(x, y) = |(x, y)|$. Hence

$$-2t|(x, y)| + t^2 \|y\|^2 \geqslant 0.$$

Rearranging and dividing by t we have, for any $t > 0$,

$$|(x, y)| \leqslant \tfrac{1}{2} t \|y\|^2.$$

Letting $t \to 0$, we infer that $(x, y) = 0$.   □

In $\mathbb{R}^3$ we can resolve a given vector into components along and perpendicular to a given direction. With the aid of the foregoing lemma we can extend this procedure to Hilbert space.

4.24 Theorem   Let M be a closed linear subspace of a Hilbert space H and let $x \in H$. There exist $y \in M$, $z \in M^\perp$ such that $x = y + z$.

Proof. As Diagram 4.3 suggests, we take y to be the closest point to x in M. That is (cf. Theorem 3.8), $y \in M$ is such that, for all $m \in M$,

$$\|x - y\| \leqslant \|x - m\|.$$

Diagram 4.2

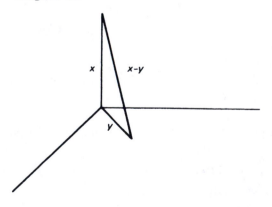

Write $z = x - y$. Then $x = y + z$. For any $m \in M$, $y + m \in M$ and so

$$\|z\| = \|x - y\| \leqslant \|x - (y + m)\|.$$

That is, for all $m \in M$,

$$\|z\| \leqslant \|z - m\|.$$

By Lemma 4.23, $z \in M^{\perp}$. $\qquad\qquad\qquad\qquad\qquad\qquad\square$

4.25 Corollary For any closed linear subspace M of a Hilbert space H,

$$(M^{\perp})^{\perp} = M.$$

Proof. It is immediate from the definition of orthogonal complements that $M \subseteq (M^{\perp})^{\perp}$. To show the reverse inclusion, consider $x \in (M^{\perp})^{\perp}$. By Theorem 4.24 we can write $x = y + z$ with $y \in M$, $z \in M^{\perp}$. Since x is orthogonal to $M^{\perp}$ (i.e. to every vector in $M^{\perp}$), $(x, z) = 0$. Thus

$$0 = (y + z, z) = (y, z) + (z, z) = \|z\|^2.$$

Hence $z = 0$ and $x = y \in M$. Thus $(M^{\perp})^{\perp} \subseteq M$. $\qquad\qquad\square$

We recall from linear algebra special terminology for the process of splitting a vector space up into smaller spaces.

4.26 Definition Let M, N be subspaces of a vector space V. We say that V is the *direct sum of M and N*, written $V = M \oplus N$, if $M \cap N = \{0\}$ and every element of V can be expressed as the sum of a member of M and a member of N. If, further, V is an inner product space and $(x, y) = 0$ for all $x \in M$, $y \in N$, then we say that V is the *orthogonal direct sum of M and N*. $\qquad\square$

Thus, for any closed subspace M of a Hilbert space H, H is the orthogonal direct sum of M and $M^{\perp}$, since clearly $M \cap M^{\perp} = \{0\}$.

Diagram 4.3

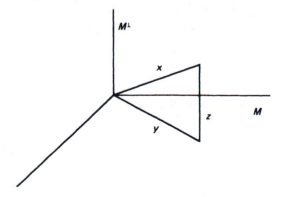

4.27 Exercise Show that $L^2(-1, 1)$ is the orthogonal direct sum of its subspaces of odd and even functions. ☐

In finite-dimensional linear algebra it is a commonplace and straightforward procedure to 'complement' a given subspace M of a vector space V: that is, to find a subspace N of V such that $V = M \oplus N$. This is still possible in infinite dimensions but, in the context of Banach spaces, the natural generalization would involve *closed* subspaces: if M is closed, can we take N to be closed too? The answer is no. There is no closed subspace N of ℓ^∞ such that $\ell^\infty = c_0 \oplus N$ (see Problem 2.7 for definitions). On the other hand, Theorem 4.24 shows that in Hilbert space the answer is yes, and a deep theorem of J. Lindenstrauss and L. Tzafriri (1971) asserts that the answer is yes *only* in Hilbert space. If a Banach space E has the property that every closed subspace has a closed complement then E is actually a Hilbert space, up to an equivalent norm.

4.5   Problems

4.1.   Find a vector in $\mathbb{C}^3$ orthogonal to $(1, 1, 1)$ and $(1, \omega, \omega^2)$, where $\omega = e^{2\pi i/3}$.

4.2. Let $e_j(z) = z^j$ for $z \in \mathbb{C}$, $j \in \mathbb{Z}$. Show that $(e_j)_{-\infty}^\infty$ is an orthonormal sequence in RL^2.

4.3. Let $f_\lambda(t) = e^{i\lambda t}$, $\lambda, t \in \mathbb{R}$. Show that $(f_\lambda)_{\lambda \in \mathbb{R}}$ is an orthonormal system in the inner product space TP of Problem 1.3. This system cannot be indexed by $\mathbb{N}$, by a theorem of Cantor, so is not an orthonormal sequence.

4.4. Let $\alpha_1, \alpha_2 \in \mathbb{D}$, the open unit disc, and let

$$e_1(z) = \frac{(1 - |\alpha_1|^2)^{1/2}}{1 - \bar{\alpha}_1 z},$$

$$e_2(z) = \frac{z - \alpha_1}{1 - \bar{\alpha}_1 z} \cdot \frac{(1 - |\alpha_2|^2)^{1/2}}{1 - \bar{\alpha}_2 z}.$$

Show that e_1, e_2 constitutes an orthonormal system in RH^2.

4.5. Let $\alpha \in \mathbb{C}$ be such that $|\alpha| \neq 1$. Find the Fourier coefficients of $f \in RL^2$, where

$$f(z) = (z - \alpha)^{-1},$$

with respect to the orthonormal sequence $(e_j)_{-\infty}^\infty$ of Problem 4.2 (consider the cases $|\alpha| < 1$ and $|\alpha| > 1$ separately).

4.6. The Gram–Schmidt process. Let $x_1, x_2, \ldots$ be a sequence of linearly independent vectors in an inner product space. Define vectors e_n inductively as follows.

$$e_1 = x_1/\|x_1\|;$$

$$f_n = x_n - \sum_{j=1}^{n-1} (x_n, e_j)e_j, \qquad n \geqslant 2;$$

$$e_n = f_n / \|f_n\|, \qquad n \geqslant 2.$$

Show that $(e_n)_1^\infty$ is an orthonormal sequence having the same closed linear span as the x_js.

4.7. The first three Legendre polynomials are

$$P_0(x) = 1, \qquad P_1(x) = x, \qquad P_2(x) = \tfrac{1}{2}(3x^2 - 1).$$

Show that the orthonormal vectors in $L^2(-1, 1)$ obtained by applying the Gram–Schmidt process to $1, x, x^2$ are scalar multiples of these.

4.8. Which point in the linear span of $(1, \omega, \omega^2)$ and $(1, \omega^2, \omega)$, where $\omega = e^{2\pi i/3}$, is nearest to $(1, -1, 1)$?

4.9. Find

$$\min_{a,b,c \,\in\, \mathbb{C}} \int_{-1}^{1} |x^3 - a - bx - cx^2|^2 \, dx$$

(use the solution of Problem 4.7).

4.10. Let H be a Hilbert space and let M be a closed linear subspace of H, so that M is a Hilbert space with respect to the restriction of the inner product of H. Let $(e_n)_1^\infty$ be a complete orthonormal sequence in M. Show that, for any $x \in H$, the best approximation to x in M is

$$y = \sum_{n=1}^{\infty} (x, e_n)e_n;$$

that is, y satisfies

$$\|x - y\| = \inf_{m \in M} \|x - m\|.$$

4.11. Prove that the standard orthonormal sequence $(e_n)_1^\infty$ is complete in ℓ^2 (e_n is the sequence with nth component equal to 1 and all others zero).

4.12. Prove that the orthonormal sequence $(e_j)_0^\infty$, where $e_j(z) = z^j$, is complete in RH^2.

[We define an orthonormal sequence in an inner product space to be complete if its closed linear span is the whole space. By Theorem 4.15, this is consistent with Definition 4.13 in the case of Hilbert space.]

4.13. Let $f \in RL^2$ have Laurent expansion

$$f(z) \sim \sum_{n=-\infty}^{\infty} a_n z^n$$

valid in an annulus containing the unit circle. Let f_+ be the best approximation to f in RH^2. Show that

$$f_+(z) = \sum_{n=0}^{\infty} a_n z^n$$

with respect to the norm of RH^2 (use Problem 4.10).

4.14. Let ℓ_Z^2 denote the Hilbert space of square-summable sequences $(x_n)_{-\infty}^\infty$ of complex numbers indexed by Z, with componentwise algebraic operations and inner product

$$((x_n), (y_n)) = \sum_{n=-\infty}^{\infty} x_n \bar{y}_n.$$

Write down an explicit isomorphism between ℓ^2 and ℓ_Z^2.

4.15. Give a definition of isomorphism of inner product spaces. Write down explicit isomorphisms between (a) RH^2 and a subspace of ℓ^2, (b) RL^2 and a subspace of ℓ_Z^2.

4.16. Is $W[-1, 1]$ the orthogonal direct sum of its subspaces of even and odd functions?

4.17. Show that in RH^2 the orthogonal complement of the space of functions

$$\{z^n f(z) : f \in RH^2\}, \qquad \text{where } n \in \mathbb{N},$$

is the space of polynomials of degree less than n.

4.18. Let y be a non-zero vector in a Hilbert space H and let

$$M = \{x \in H : (x, y) = 0\}.$$

What is $M^\perp$?

4.19. If M is a closed subspace of an inner product space V, is it necessarily true that $V = M \oplus M^\perp$? (Take $V = \ell_0$ of Example 2.7, $x = (1/n)_1^\infty \in \ell^2$ and $M = \{v \in \ell_0 : v \perp x \text{ in } \ell^2\}$).

5

Classical Fourier series

In the last chapter (and particularly in Theorem 4.14) we saw an account of orthogonal expansions in Hilbert spaces which leaves nothing to be desired – provided we know the orthonormal sequence we are using is complete. We have met one complete orthonormal sequence: the standard orthonormal sequence in ℓ^2. For this sequence, however, the theory tells us nothing that is not obvious anyway. Interesting consequences occur only with less obvious complete orthonormal sequences, so we still have some work to do before we can get any concrete information from the foregoing theory. In particular, to draw any inferences about the archetypal orthogonal expansions, classical Fourier series, we must show that Fourier's own orthonormal sequence is complete.

5.1 Theorem Let

$$e_n(x) = (2\pi)^{-1/2}\, e^{inx}, \qquad -\pi < x < \pi.$$

Then $(e_n)_{-\infty}^{\infty}$ is a complete orthonormal sequence in $L^2(-\pi, \pi)$.

Proof. As we observed in Examples 4.2, (e_n) is orthonormal. To prove completeness it suffices (by Theorem 4.15) to show that the closed linear span of the e_ns is the whole of $L^2(-\pi, \pi)$. We have to invoke a result from measure theory: the restrictions to $(-\pi, \pi)$ of the continuous 2π-periodic functions on $\mathbb{R}$ constitute a dense subspace of $L^2(-\pi, \pi)$. It thus suffices to show that every such restriction belongs to $\text{clin}\{e_n : n \in \mathbb{Z}\}$.

Consider any continuous 2π-periodic function f on $\mathbb{R}$. We wish to construct a sequence of elements of $\text{lin}\{e_n\}$ converging to f. There is an obvious candidate: let

$$f_m = \sum_{n=-m}^{m} (f, e_n)e_n.$$

Certainly $f_m \in \text{lin}\{e_n\}$, so it will suffice to prove that $f_m \to f$ (and Theorem

4.14 assures us that this must be the case if it is true that the sequence (e_n) is complete). In fact it turns out to be easier to show that

$$\frac{1}{m+1}(f_0 + f_1 + \cdots + f_m) \to f;$$

this is good enough, as the left hand side is a member of $\text{lin}\{e_n\}$.

Let

$$F_m = \frac{1}{m+1}(f_0 + f_1 + \cdots + f_m), \qquad m = 0, 1, 2, \ldots.$$

We have

$$(f, e_n) = \frac{1}{\sqrt{(2\pi)}} \int_{-\pi}^{\pi} f(x) e^{-inx} \, dx$$

and hence

$$f_m(y) = \sum_{n=-m}^{m} (f, e_n) e_n(y)$$

$$= \frac{1}{2\pi} \sum_{n=-m}^{m} \left(\int_{-\pi}^{\pi} f(x) e^{-inx} \, dx \right) e^{iny}$$

$$= \frac{1}{2\pi} \int_{-\pi}^{\pi} f(x) \sum_{n=-m}^{m} e^{in(y-x)} \, dx.$$

Thus

$$F_m(y) = \frac{1}{m+1} \sum_{j=0}^{m} f_j(y)$$

$$= \frac{1}{m+1} \sum_{j=0}^{m} \frac{1}{2\pi} \int_{-\pi}^{\pi} f(x) \sum_{n=-m}^{m} e^{in(y-x)} \, dx$$

$$= \frac{1}{2\pi} \int_{-\pi}^{\pi} f(x) \frac{1}{m+1} \sum_{j=0}^{m} \sum_{n=-j}^{j} e^{in(y-x)} \, dx.$$

Let us write

$$K_m(t) = \frac{1}{m+1} \sum_{j=0}^{m} \sum_{n=-j}^{j} e^{int}: \qquad (5.1)$$

then

$$F_m(y) = \frac{1}{2\pi} \int_{-\pi}^{\pi} f(x) K_m(y - x) \, dx. \qquad (5.2)$$

5.1   The Fejér kernel
The function K_m in (5.2) is called the Fejér kernel.

5.2 Lemma For any real t which is not an integer multiple of 2π,

$$K_m(t) = \frac{1}{m+1} \cdot \frac{\sin^2 \dfrac{(m+1)t}{2}}{\sin^2 \dfrac{t}{2}}.$$

Proof. Let us write $z = e^{it}$, $\bar{z} = e^{-it}$. For $t \notin 2\pi\mathbb{Z}$, $z \neq 1$ and

$$\sum_{n=-j}^{j} e^{int} = \bar{z}^j(1 + z + z^2 + \cdots + z^{2j})$$

$$= \bar{z}^j \frac{1 - z^{2j+1}}{1 - z} = \frac{\bar{z}^j - z^{j+1}}{1 - z}.$$

Thus, from (5.1),

$$(m+1)K_m(t) = \sum_{j=0}^{m} \frac{\bar{z}^j - z^{j+1}}{1 - z}$$

$$= \frac{1}{1-z}\left\{\frac{1 - \bar{z}^{m+1}}{1 - \bar{z}} - \frac{z(1 - z^{m+1})}{1 - z}\right\}.$$

In the second term on the right hand side multiply above and below by $\bar{z}$ and use $z\bar{z} = 1$ to obtain

$$(m+1)K_m(t) = \frac{1}{1-z}\left\{\frac{1 - \bar{z}^{m+1}}{1 - \bar{z}} + \frac{1 - z^{m+1}}{1 - \bar{z}}\right\}$$

$$= \frac{-z^{m+1} + 2 - \bar{z}^{m+1}}{|1 - z|^2}.$$

Now

$$|1 - z| = |1 - e^{it}| = |e^{-it/2} - e^{it/2}|$$
$$= 2|\sin(t/2)|,$$

and

$$z^{m+1} - 2 + \bar{z}^{m+1} = e^{i(m+1)t} - 2 + e^{-i(m+1)t}$$
$$= (e^{i(m+1)t/2} - e^{-i(m+1)t/2})^2$$
$$= \{2i \sin((m+1)t/2)\}^2.$$

Hence

$$(m+1)K_m(t) = \frac{4 \sin^2 \dfrac{(m+1)t}{2}}{4 \sin^2 \dfrac{t}{2}}. \qquad \square$$

5.3 Lemma The Fejér kernel K_m satisfies

(i) $K_m(t) \geqslant 0$ for all $t \in \mathbb{R}$, $m = 0, 1, 2, \ldots$;

(ii) $\displaystyle\int_{-\pi}^{\pi} K_m(t)\,dt = 2\pi, \qquad m = 0, 1, 2, \ldots;$

(iii) for any δ, $0 < \delta < \pi$,

$$\int_{-\pi}^{-\delta} + \int_{\delta}^{\pi} K_m(t)\,dt \to 0$$

as $m \to \infty$.

Diagram 5.1 shows the graphs of a few of the K_m. Property (iii) is becoming apparent: for large m the area under the graph of K_m is concentrated close to the point $t = 0$.

Proof. That $K_m(t) \geqslant 0$ follows from Lemma 5.2 when $t \notin 2\pi\mathbb{Z}$, and from continuity for the remaining points (K_m is a finite sum of exponentials, each of which is continuous).

Since

$$\int_{-\pi}^{\pi} e^{int}\,dt = \begin{cases} 2\pi & \text{if } n = 0 \\ 0 & \text{otherwise}, \end{cases}$$

it is obvious from the definition (5.1) of K_m that

$$\int_{-\pi}^{\pi} K_m(t)\,dt = 2\pi.$$

To prove (iii) note that if $-\pi < t < -\delta$ or $\delta < t < \pi$ then

$$\sin^2 \frac{t}{2} \geqslant \sin^2 \frac{\delta}{2} > 0,$$

and so, by Lemma 5.2,

$$0 \leqslant K_m(t) \leqslant \frac{1}{m+1} \operatorname{cosec}^2 \frac{\delta}{2}.$$

Hence

$$0 \leqslant \int_{-\pi}^{-\delta} + \int_{\delta}^{\pi} K_m(t)\,dt \leqslant \frac{2\pi \operatorname{cosec}^2 \dfrac{\delta}{2}}{m+1},$$

and for fixed $\delta > 0$, the right hand side tends to 0 as $m \to \infty$. $\square$

We are trying to show that the function F_m satisfying, for $y \in (-\pi, \pi)$,

$$F_m(y) = \frac{1}{2\pi} \int_{-\pi}^{\pi} f(x) K_m(y - x)\,dx$$

tends to f in $L^2(-\pi, \pi)$ as $m \to \infty$. Before continuing the formal proof let us try to gain a feel for the reason that this is so. We must develop an intuition for the properties of the function $k * f$, defined by

$$k * f(y) = \int_{-\pi}^{\pi} f(x) k(y - x)\,dx,$$

in terms of the properties of f and k (we take both f and k to be piecewise continuous 2π-periodic functions on $\mathbb{R}$). $k * f$ is called the *convolution* of k and f. Convolution is one of the basic operations of analysis: it turns up frequently in both theory and applications.

Suppose that δ is a small positive number and that k is the function which is zero on $(-\pi, -\delta) \cup (\delta, \pi)$ and is constant on $(-\delta, \delta)$, taking a value such that the total area under the graph of k over $(-\pi, \pi)$ is 1. Thus k takes the value $1/2\delta$ on $(-\delta, \delta)$. It is continued to a function on $\mathbb{R}$ in such a way as to be 2π-periodic. For any y,

$$k * f(y) = \int_{-\pi}^{\pi} f(x)k(y - x)\,dx$$

$$= \int_{y-\delta}^{y+\delta} \frac{1}{2\delta} f(x)\,dx.$$

Diagram 5.1   Fejér's kernel

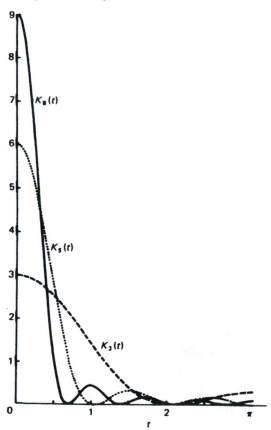

In other words, convolving f with k replaces the value of f at y by its average over the interval $(y - \delta, y + \delta)$. One can think of this as a kind of smoothing process: indeed, it is sometimes used as such in signal processing. If a function f as measured is contaminated by random errors ('noise') one could hope to improve matters by averaging them out locally, that is, by forming $k * f$ with a small value of δ. It is to be expected that if f is a continuous function and δ is small then $k * f$ will be close to f, and indeed the remainder of the proof of Theorem 5.1 consists essentially in showing that this is so. It is true that the kernel K_m we are using differs from k, but Diagram 5.1 and, more formally, Lemma 5.3 show that for large m K_m becomes rather like k, so that we can think of $K_m * f(y)$ as a weighted average of the values of f close to y.

Conclusion of proof of 5.1. Consider a fixed point $y \in [-\pi, \pi]$. Make the substitution $t = y - x$ in Lemma 5.3(ii) to obtain

$$\int_{y-\pi}^{y+\pi} K_m(y-x)\,dx = 2\pi.$$

Multiply by $f(y)/2\pi$:

$$f(y) = \frac{1}{2\pi} \int_{y-\pi}^{y+\pi} f(y) K_m(y-x)\,dx. \tag{5.3}$$

Now K_m is easily seen from its definition (equation (5.1)) to be 2π-periodic, and so the same is true for the function $x \mapsto f(x) K_m(y-x)$. The integral of this function thus takes the same value over any complete period, and in particular is the same over $[-\pi, \pi]$ and $[y - \pi, y + \pi]$. We can therefore re-write equation (5.2) in the form

$$F_m(y) = \frac{1}{2\pi} \int_{y-\pi}^{y+\pi} f(x) K_m(y-x)\,dx.$$

On subtracting equation (5.3) we find

$$F_m(y) - f(y) = \frac{1}{2\pi} \int_{y-\pi}^{y+\pi} [f(x) - f(y)] K_m(y-x)\,dx. \tag{5.4}$$

This integral is small for large m because, when x is close to y, $f(x) - f(y)$ is small, while, when x is distant from y, $K_m(y-x)$ is small. We make this argument precise. We have, for any $y \in \mathbb{R}$, $0 < \delta < \pi$,

$$|F_m(y) - f(y)| \leq \frac{1}{2\pi} \int_{y-\pi}^{y+\pi} |f(x) - f(y)| K_m(y-x)\,dx$$

$$= \frac{1}{2\pi} \left[\int_{y-\pi}^{y-\delta} + \int_{y-\delta}^{y+\delta} + \int_{y+\delta}^{y+\pi} \right] |f(x) - f(y)| K_m(y-x)\,dx. \tag{5.5}$$

Let $\varepsilon > 0$. We wish to find $m_0 \in \mathbb{N}$ such that $m \geqslant m_0$ implies that $|F_m(y) - f(y)| < \varepsilon$ for all $y \in \mathbb{R}$.

Since f is continuous on $[-\pi, \pi]$ it is also bounded, so that there exists $M > 0$ such that

$$|f(x)| \leqslant M, \quad \text{all } x \in \mathbb{R}.$$

A continuous function on a compact interval is uniformly continuous (see, for example, Binmore, 1981, Book 2, p. 147). Applying this theorem to f on the interval $[-2\pi, 2\pi]$, we infer that there exists δ, $0 < \delta < \pi$, such that $-\pi \leqslant y \leqslant \pi$, $|x - y| < \delta$ imply

$$|f(x) - f(y)| < \frac{\varepsilon}{2}.$$

By Lemma 5.3(iii) there exists $m_0 \in \mathbb{N}$ such that $m \geqslant m_0$ implies

$$\int_{-\pi}^{-\delta} + \int_{\delta}^{\pi} K_m(t)\, \mathrm{d}t < \frac{\pi\varepsilon}{2M}.$$

Making the change of variable $t = y - x$, we deduce that

$$\int_{y-\pi}^{y-\delta} + \int_{y+\delta}^{y+\pi} K_m(y - x)\, \mathrm{d}x < \frac{\pi\varepsilon}{2M}.$$

Since $|f(x) - f(y)| \leqslant 2M$ for any x, y, we deduce that when $m \geqslant m_0$,

$$\frac{1}{2\pi} \int_{y-\pi}^{y-\delta} + \int_{y+\delta}^{y+\pi} |f(x) - f(y)| K_m(y - x)\, \mathrm{d}x$$

$$< \frac{1}{2\pi} \cdot 2M \cdot \frac{\pi\varepsilon}{2M} = \frac{\varepsilon}{2}. \tag{5.6}$$

On the other hand, with δ as chosen and any $m \in \mathbb{N}$, $|f(x) - f(y)| < \varepsilon/2$ for $x \in (y - \delta, y + \delta)$, so that

$$\frac{1}{2\pi} \int_{y-\delta}^{y+\delta} |f(x) - f(y)| K_m(y - x)\, \mathrm{d}x \leqslant \frac{\varepsilon}{2} \cdot \frac{1}{2\pi} \int_{y-\delta}^{y+\delta} K_m(y - x)\, \mathrm{d}x$$

$$\leqslant \frac{\varepsilon}{2} \cdot \frac{1}{2\pi} \int_{y-\pi}^{y+\pi} K_m(y - x)\, \mathrm{d}x$$

$$= \frac{\varepsilon}{2}.$$

On substituting this and the estimate (5.6) into (5.5) we conclude that, for all $y \in [-\pi, \pi]$ and $m \geqslant m_0$,

$$|F_m(y) - f(y)| < \frac{\varepsilon}{2} + \frac{\varepsilon}{2} = \varepsilon.$$

We have now shown that $F_m \to f$ uniformly on $[-\pi, \pi]$ as $m \to \infty$; in other words,

$$\|F_m - f\|_\infty \to 0,$$

where $\|\cdot\|_\infty$ is the supremum norm on $[-\pi, \pi]$. This implies that

$$\|F_m - f\| \to 0,$$

where $\|\cdot\|$ is the norm of $L^2(-\pi, \pi)$. For if $g \in C[-\pi, \pi]$,

$$\|g\|^2 = \int_{-\pi}^{\pi} |g(x)|^2 \, dx \leqslant \|g\|_\infty^2 \int_{-\pi}^{\pi} dx$$
$$= 2\pi \|g\|_\infty^2.$$

Thus

$$0 \leqslant \|F_m - f\| \leqslant \sqrt{(2\pi)} \|F_m - f\|_\infty \to 0.$$

We have now shown that if f is the restriction to $(-\pi, \pi)$ of a continuous 2π-periodic function on $\mathbb{R}$ then $f \in \text{clin}\{e_n : n \in \mathbb{Z}\}$, as required.   □

As a reward for the considerable effort of establishing the completeness of the trigonometric sequence, we now have something to which we can apply the general theory of orthogonal expansions contained in Chapter 4. In particular, Theorem 4.14 tells us that every L^2 function is the sum of its Fourier series in the L^2 sense. More precisely:

5.4 Corollary   Let $f \in L^2(-\pi, \pi)$ have Fourier series

$$f(x) \sim \sum_{-\infty}^{\infty} c_n e^{inx}.$$

Then

$$\left\| f(x) - \sum_{n=-m}^{m} c_n e^{inx} \right\| \to 0$$

in $L^2(-\pi, \pi)$ as $m \to \infty$.   □

5.2   Fejér's theorem

Although Corollary 5.4 gives us valuable information about Fourier series, it falls short of telling us what we might want to know for applications. Suppose that we wish to know the value of a function f at a particular point, and are able to calculate the Fourier series of f – say from a differential equation. Can we obtain the desired value as the limit of the Fourier series of f at the point in question? The last Corollary tells us only about L^2 convergence and, as we saw in Caution 4.12, this does not imply convergence at each point. However, in the course of proving the completeness of the trigonometric sequence we proved something stronger, which *does* tell us about convergence of Fourier series at a point.

5.5 *Theorem* (Fejér)   Let f be a continuous 2π-periodic function on $\mathbb{R}$, let $s_n(f)$ be the nth partial sum of the Fourier series of f and let $\sigma_n(f)$ be the arithmetic mean of $s_0(f), \ldots, s_n(f)$. Then $\sigma_n(f) \to f$ uniformly on $\mathbb{R}$ as $n \to \infty$. $\square$

Let us spell out the precise meaning of this statement. Suppose that the Fourier series of f is

$$f(x) \sim \sum_{n=-\infty}^{\infty} c_n e^{inx};$$

we write

$$s_n(f, x) = \sum_{j=-n}^{n} c_j e^{ijx}$$

and

$$\sigma_n(f, x) = \frac{1}{n+1} \sum_{j=0}^{n} s_j(f, x).$$

Fejér's theorem asserts that, for continuous 2π-periodic functions f,

$$\lim_{n \to \infty} \sigma_n(f, x) = f(x)$$

uniformly in x as $n \to \infty$. In other words, for all $\varepsilon > 0$ there exists $n_0 \in \mathbb{N}$ such that $n \geq n_0$ implies $|f(x) - \sigma_n(f, x)| < \varepsilon$ for all $x \in \mathbb{R}$.

In the notation of the proof of Theorem 5.1,

$$s_n(f, x) = f_n(x), \qquad \sigma_n(f, x) = F_n(x),$$

and we did show that $F_n \to f$ uniformly on $\mathbb{R}$.

Fejér's theorem provides a procedure for calculating $f(x)$ in the hypothetical situation above, at least if f is known to be continuous. In fact it is enough for f to be continuous at the point in question: see Problem 5.6. One obtains $f(x)$ as the limit of the sequence $\sigma_n(f, x)$. One can hardly help wondering whether it would not do to use the sequence $s_n(f, x)$ instead of $\sigma_n(f, x)$. This would make the statement of the theorem neater, though it does not make much difference as far as computation is concerned. The answer is no: there are continuous functions whose Fourier series fail to converge at some point (Hardy and Rogosinski, 1965, p. 50). However, the set of points at which the Fourier series of any function in $L^2(-\pi, \pi)$ (and *a fortiori* any continuous 2π-periodic function) diverges cannot be too large. For many years it was one of the best-known unsolved problems in analysis to determine whether such a set of divergence had to be of measure zero. That it had was called the Lusin conjecture; it was proved to be true in 1966 by L. Carleson.

If the Fourier series of a continuous function f does converge at a point x then the sum of the series is actually $f(x)$: see Problem 5.7.

5.3   Parseval's formula

5.6   *Theorem* (Parseval's formula)   Let $f, g \in L^2(-\pi, \pi)$ have Fourier series

$$f(x) \sim \sum c_n e^{inx}, \qquad g(x) \sim \sum d_n e^{inx}.$$

Then

$$\frac{1}{2\pi} \int_{-\pi}^{\pi} f(x)\overline{g(x)}\, dx = \sum_{n=-\infty}^{\infty} c_n \bar{d}_n.$$

Proof. Let us denote by ℓ_Z^2 the Hilbert space of square summable sequences $(\xi_n)_{n \in Z}$ of complex numbers indexed by $\mathbb{Z}$, with componentwise addition and scalar multiplication and inner product

$$((\xi_n), (\eta_n)) = \sum_{n=-\infty}^{\infty} \xi_n \bar{\eta}_n.$$

ℓ_Z^2 is a 'renumbering' of ℓ^2. In the notation of Theorem 5.1, $(e_n)_{n \in Z}$ is a complete orthonormal sequence in $L^2(-\pi, \pi)$, and hence the mapping $U: L^2(-\pi, \pi) \to \ell_Z^2$ defined by $Uf = (\xi_n)_{n \in Z}, \xi_n = (f, e_n)$, is a unitary operator (see Theorem 4.19). It follows that $(f, g) = (Uf, Ug)$. Now

$$c_n = \frac{1}{2\pi} \int_{-\pi}^{\pi} f(x) e^{-inx}\, dx = \frac{1}{\sqrt{(2\pi)}} (f, e_n),$$

and hence

$$Uf = \sqrt{(2\pi)}(c_n)_{n \in Z}, \qquad Ug = \sqrt{(2\pi)}(d_n)_{n \in Z}.$$

Thus

$$\int_{-\pi}^{\pi} f(x)\overline{g(x)}\, dx = (f, g)$$

$$= (Uf, Ug)$$

$$= 2\pi \sum_{n=-\infty}^{\infty} c_n \bar{d}_n. \qquad \square$$

5.7 *Corollary*   If $f \in L^2(-\pi, \pi)$ and

$$f(x) \sim \sum c_n e^{inx}$$

then

$$\frac{1}{2\pi} \int_{-\pi}^{\pi} |f(x)|^2\, dx = \sum_{n=-\infty}^{\infty} |c_n|^2. \qquad \square$$

5.4   Weierstrass' approximation theorem

A by-product of our labours is a major classical result about approximating functions by polynomials.

5.8 Weierstrass' approximation theorem   Let $a < b$ in $\mathbb{R}$. For any continuous $\mathbb{C}$-valued function f on $[a, b]$ and any $\varepsilon > 0$ there exists a complex polynomial p such that

$$|f(x) - p(x)| < \varepsilon, \qquad a \leqslant x \leqslant b.$$

Proof. By changing to the independent variable

$$\xi = \frac{\pi}{b - a}\left(x - \frac{a + b}{2}\right)$$

we can arrange that $[a, b]$ becomes $[-\pi, \pi]$. We wish to show that the space $\wp$ of polynomial functions on $[-\pi, \pi]$ is dense in $C[-\pi, \pi]$, with respect to the supremum norm. Denote the closure of $\wp$ by clos $\wp$. Now the Taylor series of e^z converges to e^z uniformly on compact subsets of $\mathbb{C}$: hence the function e^{inx} is in clos $\wp$ for each $n \in \mathbb{Z}$. clos $\wp$ is a subspace of $C[-\pi, \pi]$, and so it contains all linear combinations of the functions e^{inx}, and so in particular it contains the function $\sigma_n(g)$, the arithmetic mean of the first $n + 1$ partial sums of the Fourier series of g, for any continuous 2π-periodic function g. By Fejér's theorem $\sigma_n(g) \to g$ with respect to the supremum norm: hence clos $\wp$ contains all continuous 2π-periodic functions on $[-\pi, \pi]$. It is possible that $f(-\pi) \neq f(\pi)$, but we can easily choose a complex number c such that $f(x) + cx$ is 2π-periodic (i.e. has the same value at π and $-\pi$). Then $f(x) + cx \in$ clos $\wp$, and hence $f \in$ clos $\wp$. $\square$

5.5   Problems

5.1.   Find the Fourier series of the following functions on $[-\pi, \pi]$ and use Corollary 5.7 to deduce the stated formulae.

(a) $f(x) = x$; $\displaystyle\sum_{n=1}^{\infty} \frac{1}{n^2} = \frac{\pi^2}{6}$.

(b) $f(x) = x^2$; $\displaystyle\sum_{n=1}^{\infty} \frac{1}{n^4} = \frac{\pi^4}{90}$.

(c) $f(x) = e^{sx}$; $\displaystyle\sum_{n=-\infty}^{\infty} \frac{1}{n^2 + s^2} = \frac{\pi}{s}\coth \pi s$.

5.2. Let $f_0(x) = (2\pi)^{-1/2}$, $f_n(x) = \pi^{-1/2}\cos nx$, $g_n(x) = \pi^{-1/2}\sin nx$ for $n \geqslant 1$. Show (using Theorem 5.1) that $(f_0, g_1, f_1, g_2, f_2, \ldots)$ is a complete orthonormal sequence in $L^2(-\pi, \pi)$.

5.3. Show that $(e^{2\pi inx})_{n=-\infty}^{\infty}$ is a complete orthonormal sequence in $L^2(0, 1)$.

5.4. Let

$$g_n(x) = \frac{e^{inx}}{\sqrt{\{2\pi(1 + n^2)\}}}, \qquad n \in \mathbb{Z}, \; -\pi \leqslant x \leqslant \pi.$$

Show that $(g_n)_{-\infty}^{\infty}$ is an orthonormal sequence in $W[-\pi, \pi]$.

Let
$$W_0 = \{f \in W[-\pi, \pi] : f(\pi) = f(-\pi)\}.$$
Show that sinh is orthogonal to every function in W_0. Deduce that $(g_n)_{-\infty}^{\infty}$ is not a complete orthonormal sequence in $W[-\pi, \pi]$.

5.5. Show that if $f \in L^2(-\pi, \pi)$ has Fourier series
$$f(x) \sim a_0 + \sum_{n=1}^{\infty} a_n \cos nx + b_n \sin nx$$
then
$$\frac{1}{2\pi} \int_{-\pi}^{\pi} |f(x)|^2 \, dx = |a_0|^2 + \frac{1}{2} \sum_{n=1}^{\infty} |a_n|^2 + |b_n|^2.$$

5.6. Modify the proof of Fejér's theorem to show that it holds 'pointwise'. That is, let $f \in L^2(-\pi, \pi)$ be bounded, let
$$f(x) \sim \sum_{-\infty}^{\infty} c_n e^{inx}$$
and let f be continuous at $y \in (-\pi, \pi)$. Let
$$s_m(f, x) = \sum_{n=-m}^{m} c_n e^{inx},$$
$$\sigma_m(f, x) = \frac{1}{m+1} \sum_{j=0}^{m} s_j(f, x).$$
Prove that $\sigma_m(f, y) \to f(y)$ as $m \to \infty$.

5.7. Let $s_n \in \mathbb{C}, n = 0, 1, \ldots$, and let σ_n be the arithmetic mean of $s_0, \ldots, s_n$. Show that if $s_n \to s$ as $n \to \infty$ then $\sigma_n \to s$ as $n \to \infty$.

Assuming the result of Problem 5.6 prove that if $f \in L^2(-\pi, \pi)$ is continuous at $y \in (-\pi, \pi)$, and if the Fourier series of f does converge at y then its limit is $f(y)$.

5.8. Let $-\infty < a < b < \infty$. Show that there is an orthonormal sequence $(f_n)_0^{\infty}$ in $L^2(a, b)$ such that f_n is a polynomial of degree n. Show further that $(f_n)_0^{\infty}$ is complete in $L^2(a, b)$ and that each f_n is uniquely determined up to multiplication by a scalar of unit modulus.

5.9. Let $\varepsilon > 0$ and let $f \in W_0$, in the notation of Problem 5.4. Show that the zeroth (classical) Fourier coefficient of f' is zero. Deduce that there exists $g \in \lim\{e^{inx} : n \in \mathbb{Z}, n \neq 0\}$ such that
$$\|f' - g\|_{L^2} < \frac{\varepsilon}{\sqrt{(1 + 4\pi^2)}}.$$
Let
$$G(x) = f(-\pi) + \int_{-\pi}^{x} g(\tau) \, d\tau, \qquad -\pi \leqslant x \leqslant \pi.$$

Show that $G \in \mathrm{lin}\{e^{inx} : n \in \mathbb{Z}\}$ and that

$$\|f - G\|_{W[-\pi,\pi]} < \varepsilon.$$

Deduce that $(g_n)^{\infty}_{-\infty}$, together with a suitable scalar multiple of sinh, constitutes a complete orthonormal system in $W[-\pi, \pi]$.

6

Dual spaces

The set of linear functionals on an n-dimensional vector space is itself an n-dimensional vector space with respect to the obvious algebraic operations, and relations in the original vector space can sometimes profitably be dualized – that is, converted to statements applying to the space of functionals. Phenomena in the most diverse applications of mathematics, ranging from economics to combinatorics and the design of experiments, can be viewed as manifestations of this principle. Indeed the idea of duality pervades mathematics. It plays a central role in functional analysis especially.

It is easy to write down the general linear functional on a finite-dimensional space. In infinite dimensions things are trickier. In the first place, it appears that the set of *all* linear functionals on a normed space is too big to be useful: one has to restrict attention to *continuous* linear functionals. Even these are quite intangible objects in general but, for most of the commonly-encountered normed spaces, descriptions of the continuous linear functionals are listed in standard reference works such as Dunford an 1 Schwartz (1957). We give a simple example to illustrate the nature of such descriptions (Example 6.7 below).

One overwhelming advantage of Hilbert space over Banach spaces is that the general linear functional *can* be written down: it is simply of the form $(\cdot, y)$ for some element y of the space. This is the Riesz–Fréchet theorem (6.8 below). Essentially, a Hilbert space is its own dual. This fact makes duality in Hilbert spaces much closer to finite-dimensional duality, and for that reason it is a particularly powerful tool.

6.1 Definition   A *linear functional* on a vector space E over a field k is a mapping $f: E \to k$ which satisfies

$$f(\lambda x + \mu y) = \lambda f(x) + \mu f(y)$$

for all $\lambda, \mu \in k$ and $x, y \in E$.   □

6.2 Examples   (i) Define $F: \mathbb{C}^n \to \mathbb{C}$ by

$$F(x_1, \ldots, x_n) = c_1 x_1 + c_2 x_2 + \cdots + c_n x_n$$

where $c_1, \ldots, c_n \in \mathbb{C}$.

(ii) Define $F_c: \ell^1 \to \mathbb{C}$, where ℓ^1 is the sequence space defined in Problem 2.2, by

$$F_c((x_n)_{n \in \mathbb{N}}) = \sum_{n=1}^{\infty} c_n x_n$$

where $c = (c_n)_{n \in \mathbb{N}} \in \ell^{\infty}$ (see Problem 2.7). To say that $(c_n) \in \ell^{\infty}$ simply means that the c_n are bounded. For $(x_n) \in \ell^1$, $\sum |x_n|$ is finite, and hence $\sum c_n x_n$ is an absolutely convergent series. Thus F_c is well defined on ℓ^1.

(iii) Define $F: C[0, 1] \to \mathbb{C}$ by

$$F(x) = \int_0^1 x(t) \, d\alpha(t)$$

where α is a function of bounded variation on $[0, 1]$.

(iv) Let H be a Hilbert space and let $y \in H$. Define $F_y: H \to \mathbb{C}$ by

$$F_y(x) = (x, y).$$ □

It is easy to check that each of these examples is a linear functional. In fact they are also all continuous. The following simple observation is useful for checking this.

6.3 Theorem   Let F be a linear functional on a normed space $(E, \| \cdot \|)$. The following are equivalent:

(i) F is continuous;

(ii) F is continuous at 0;

(iii) $\sup\{|F(x)| : x \in E, \|x\| \leqslant 1\} < \infty$.

Thus F is continuous if and only if it is bounded on the unit ball of E.

Proof. It is trivial that (i) implies (ii). Suppose (ii): then there exists $\delta > 0$ such that $\|x\| < \delta$ implies $|F(x)| < 1$. Now for any $x \in E$ such that $\|x\| \leqslant 1$ we have $\|\delta x/2\| < \delta$ and hence $|F(\delta x/2)| < 1$. Thus $|F(x)| < 2/\delta$, and so (iii) holds.

Finally, suppose (iii) holds, and denote the finite supremum in (iii) by M. For any pair x, y of distinct vectors in E, $(x - y)/\|x - y\|$ is a unit vector and hence

$$\left| F\left(\frac{x - y}{\|x - y\|} \right) \right| \leqslant M.$$

Thus

$$|F(x) - F(y)| \leqslant M \|x - y\|,$$

from which it is clear that F is continuous.   □

6.4 *Exercise* Define a linear functional F on the space $C^1[0, 1]$ of continuously differentiable functions on $[0, 1]$ by

$$F(x) = x'(1).$$

Show that F is discontinuous with respect to the supremum norm $\| \cdot \|_\infty$.

6.5 *Theorem* The set E^* of all continuous linear functionals on the normed space $(E, \| \cdot \|)$ is itself a Banach space with respect to pointwise algebraic operations and norm

$$\|F\| = \sup_{x \in E, \|x\| \leqslant 1} |F(x)|.$$

Proof. E^* is clearly a vector space over the same field as E. By Theorem 6.3, $\|F\|$ is a real number for each $F \in E^*$, and it is easy to check that $\| \cdot \|$ is a norm on E^*. It remains to show that E^* is complete: note that this is so whether or not E itself is complete.

Let (F_n) be a Cauchy sequence in E^*, so that $\|F_n - F_m\| \to 0$ as $n, m \to \infty$. Hence, for any $x \in E$, $|F_n(x) - F_m(x)| \to 0$ as $n, m \to \infty$; that is, $(F_n(x))_n$ is a Cauchy sequence of scalars. Denote the limit of this sequence by $F(x)$. It is easy to see that F is a linear functional on E: we must show that $F \in E^*$ and $F_n \to F$ with respect to the norm of E^*.

Let $\varepsilon > 0$ and pick $n_0 \in \mathbb{N}$ such that $m, n \geqslant n_0$ implies that $\|F_m - F_n\| < \varepsilon$. Then for any $x \in E$ such that $\|x\| \leqslant 1$ and for all $m, n \geqslant n_0$,

$$|F_m(x) - F_n(x)| < \varepsilon.$$

Letting $m \to \infty$ we infer that for all $n \geqslant n_0$ and all x such that $\|x\| \leqslant 1$,

$$|F(x) - F_n(x)| \leqslant \varepsilon. \tag{6.1}$$

Since both $F - F_{n_0}$ and F_{n_0} are bounded on the closed unit ball of E, F is also, and so F is continuous by Theorem 6.3. Thus $F \in E^*$. And (6.1) states that $\|F - F_n\| \leqslant \varepsilon$ whenever $n \geqslant n_0$. Thus $F_n \to F$ in E^*, and so E^* is complete. $\qquad \square$

$(E^*, \| \cdot \|)$ is called the *dual space*, or sometimes the *Banach dual* of E.

Let us note the following easy consequence of the definition of the dual norm.

6.6 *Theorem* For any vector x in a normed space E and any continuous linear functional F on E,

$$|F(x)| \leqslant \|F\| \|x\|.$$

Proof. We can suppose $x \neq 0$. Then $x/\|x\|$ is a unit vector, and so

$$\|F\| \geqslant |F(x/\|x\|)| = \frac{|F(x)|}{\|x\|}. \qquad \square$$

Here now is a sample result describing the dual space of a particular Banach space.

6.7 Example $(\ell^1)^*$ can be identified with ℓ^∞. That is, there is a mapping $T: \ell^\infty \to (\ell^1)^*$ which is at once an isomorphism of vector spaces and an isometry (norm-preserving mapping). Indeed, let $Tc = F_c$ in the notation of Example 6.2(ii). T is evidently linear; let us show that it is surjective. Pick any $g \in (\ell^1)^*$: we must find $c \in \ell^\infty$ such that $F_c = g$. Let e_n denote the 'nth standard basis vector' in ℓ^1, let $c_n = g(e_n)$ and let $c = (c_n)_{n \in \mathbb{N}}$ [this is the only possible candidate for c since $F_c(e_n) = c_n$]. Since e_n is a unit vector in ℓ^1,

$$|c_n| = |g(e_n)| \leqslant \|g\|.$$

Thus c is a bounded sequence, i.e. $c \in \ell^\infty$, and

$$\|c\|_\infty = \sup_n |c_n| \leqslant \|g\|. \tag{6.2}$$

To show that $F_c = g$, consider any $x = (x_n) \in \ell^1$ and write $x^k = \sum_{n=1}^k x_n e_n$. Then $x^k \to x$ in ℓ^1 as $k \to \infty$. We have

$$F_c(x^k) - g(x^k) = \sum_{n=1}^k c_n x_n - \sum_{n=1}^k x_n g(e_n) = 0,$$

and hence, on letting $k \to \infty$, $F_c(x) = g(x)$.

It remains to show that T preserves norms. We have already seen in (6.2) above that $\|c\|_\infty \leqslant \|F_c\|$. And, for any $x \in \ell^1$,

$$|F_c(x)| = \left| \sum_{n=1}^\infty c_n x_n \right| \leqslant \left(\sup_n |c_n| \right) \sum |x_n|$$
$$= \|c\|_\infty \|x\|_1.$$

Hence $\|F_c\| \leqslant \|c\|_\infty$. Thus $\|F_c\| = \|c\|_\infty$, and so T preserves norms. ☐

One might expect, by analogy with the finite-dimensional theory, that the dual space of ℓ^∞ could in turn be identified with ℓ^1. It is indeed true that an element $x \in \ell^1$ determines a continuous linear functional on ℓ^∞, to wit the mapping $c \mapsto F_c(x)$, and moreover the norm of this functional coincides with $\|x\|_1$. This only shows, though, that ℓ^1 can be identified with a *subspace* of $(\ell^\infty)^*$, and in fact $(\ell^\infty)^*$ is a much larger space.

6.1   The Riesz–Fréchet theorem

6.8   *Theorem* (Riesz–Fréchet)   Let H be a Hilbert space and let F be a continuous linear functional on H. There exists a unique $y \in H$ such that

$$F(x) = (x, y)$$

for all $x \in H$. Moreover $\|y\| = \|F\|$.

Proof. Note first that there can be at most one element $y \in H$ with the property described, for if

$$(x, y) = F(x) = (x, y')$$

for all $x \in H$ then $y = y'$ by Theorem 1.5(iv).

If F is the zero functional the statement is true: take $y = 0$. Otherwise

$$M = \operatorname{Ker} F = \{x \in H : F(x) = 0\}$$

is a proper closed subspace of H. By Theorem 4.24, $H = M \oplus M^\perp$ and hence $M^\perp \neq \{0\}$. Pick a non-zero vector $z \in M^\perp$. By multiplying z by a suitable scalar we can arrange that $F(z) = 1$ [note that if there *is* an element y as described it must certainly belong to $M^\perp$]. For any $x \in H$ we have

$$x = (x - F(x)z) + F(x)z.$$

The first term on the right hand side belongs to M and the second to $M^\perp$. Taking the inner product of both sides with z and using the fact that $z \perp M$, we obtain

$$(x, z) = (F(x)z, z) = F(x)\|z\|^2$$

for all $x \in H$. Let $y = z/\|z\|^2$: then

$$(x, y) = (x, z)/\|z\|^2 = F(x)$$

for all $x \in H$; this proves the first statement.

If $\|x\| \leqslant 1$ then, by the Cauchy–Schwarz inequality,

$$|F(x)| = |(x, y)| \leqslant \|x\| \|y\|$$
$$\leqslant \|y\|,$$

and therefore $\|F\| \leqslant \|y\|$. On the other hand, $x = y/\|y\|$ is a unit vector (we are still supposing $F \neq 0$, which clearly entails $y \neq 0$), and therefore

$$\|F\| \geqslant |F(x)| = |F(y)|/\|y\| = |(y, y)|/\|y\|$$
$$= \|y\|.$$

Thus $\|F\| = \|y\|$.   □

The theorem shows that, for any Hilbert space H, there is a surjective and norm-preserving mapping $T : H \to H^*$ given by

$$Ty = (\cdot, y).$$

T is also *conjugate linear*; that is,

$$T(\lambda y + \mu z) = \bar{\lambda} Ty + \bar{\mu} Tz.$$

In view of the existence of such a T, Hilbert spaces are sometimes said to be 'self-dual'.

Theorem 6.8 tells us just about all there is to know about linear functionals on a Hilbert space. For normed spaces the issues are very different. Consider, for example, a simple question: if x, y are distinct elements of a normed space E, is there a continuous linear functional F on E such that $F(x) \neq F(y)$? If E is a Hilbert space we can immediately say yes, for we may take $F = (\cdot, x - y)$, but for general E the answer is not obvious. Might it even be that for some E there are no continuous linear functionals other than the zero functional? If this were so then duality would not provide useful techniques for the study of normed spaces. In fact there are always plenty of continuous linear functionals on any normed space, and there is always an $F \in E^*$ such that $F(x) \neq F(y)$: this is a consequence of the Hahn–Banach theorem, which is proved in most introductory texts on functional analysis (see, for example, Rudin, 1973). The methods of proof are quite different from anything in this book: the commonest one is an element-by-element construction. This and all other proofs require one of the higher axioms of set theory, such as Zorn's lemma.

6.2   Problems

6.1.   Let $-\infty < a < c < b < \infty$. Show that the linear functional F on $C[a, b]$ defined by

$$F(x) = x(c), \qquad x \in C[a, b],$$

is continuous with respect to the supremum norm, but not with respect to the $L^2(a, b)$ norm (restricted to $C[a, b]$).

Does it make sense to define a functional G on $L^2(a, b)$ by

$$G(x) = x(c)?$$

6.2. Define the nth *coefficient functional* on the space $\wp$ of polynomial functions on $[0, 1]$ to be the functional C_n given by

$$C_n(p) = a_n$$

if

$$p(t) = \sum_{j=0}^{m} a_j t^j, \qquad 0 \leqslant t \leqslant 1,$$

with the understanding that $C_n(p) = 0$ if $m < n$. Show that C_n is discontinuous with respect to the supremum norm (consider $p_k(t) = (1 - t)^k \Big/ \binom{k}{n}$).

6.3. Define the nth coefficient functional C_n on RH^2 by

$$C_n(f) = a_n$$

if $f(z)$ has power series expansion $\sum_0^\infty a_j z^j$ in $\mathbb{D}$. Show that C_n is continuous on RH^2.

6.4. What is the norm of the trace as a linear functional on $\mathbb{C}^{n \times n}$ (cf. Exercise 1.4)?

6.5. Find the norm of the linear functional

$$F(x) = \int_0^1 tx(t)\,dt$$

on $(C[0, 1], \|\cdot\|_\infty)$. Find also an element of $C[0, 1]$ at which F attains its norm: that is, an element x of unit norm such that

$$|F(x)| = \|F\|.$$

6.6. Let

$$E = \{x \in C[0, 1] : x(1) = 0\},$$

and let G be the restriction to E of the linear functional F of the preceding problem. Show that $\|G\| = \|F\|$, but that G does not attain its norm on $(E, \|\cdot\|_\infty)$.

6.7. Let F be a non-zero continuous linear functional on a Banach space E and let

$$M = \{x \in E : F(x) = 1\}.$$

Prove that M is closed, convex and non-empty. Show further that

$$\inf_{x \in M} \|x\| = \frac{1}{\|F\|},$$

and that, if F does not attain its norm on E then there is no closest point to the origin in M (i.e. the infimum above is not attained).

6.8. Let F be a linear functional on a normed space E and let

$$\operatorname{Ker} F = \{x \in E : F(x) = 0\}.$$

Show that $\operatorname{Ker} F$ has co-dimension 1 in E (definition in Problem 2.12). Deduce that if V is a subspace of E containing $\operatorname{Ker} F$ then either $V = E$ or $V = \operatorname{Ker} F$. Hence show that if F is discontinuous then $\operatorname{Ker} F$ is dense in E. Conclude that F is continuous if and only if $\operatorname{Ker} F$ is closed.

6.9. Prove that $(c_0)^*$ can be identified with ℓ^1: that is, there is a mapping $T : \ell^1 \to (c_0)^*$ which is an isomorphism of vector spaces and which preserves norms. Why does your proof not show that $(\ell^\infty)^*$ can be identified with ℓ^1?

6.10. Let $0 \leqslant \alpha \leqslant 1$. Does the formula

$$F(x) = x(\alpha)$$

define a bounded linear functional on $W[0, 1]$? If so, what is $\|F\|$?

Answer this by carrying out the following steps.

(i) Introduce the space $\tilde{W}[0, 1]$ of continuous functions on $[0, 1]$ which are piecewise continuously differentiable (x is piecewise continuously differentiable on $[0, 1]$ if there is a partition $0 = t_0 < t_1 < t_2 < \cdots < t_n = 1$ such that x is continuously differentiable on each $[t_j, t_{j+1}]$: the right and left derivatives of x at each t_j must exist but may differ). $\tilde{W}[0, 1]$ is an inner product space with respect to

$$(x, y) = \int_0^1 x(t)\overline{y(t)} + x'(t)\overline{y'(t)}\, dt.$$

(ii) Let

$$h_\alpha(t) = \begin{cases} \text{cosech } 1 \cosh(1 - \alpha) \cosh t & \text{if } 0 \leqslant t \leqslant \alpha, \\ \text{cosech } 1 \cosh \alpha \cosh(1 - t) & \alpha < t \leqslant 1. \end{cases}$$

Then $h_\alpha \in \tilde{W}[0, 1]$ and

$$(x, h_\alpha) = x(\alpha).$$

(iii) F is bounded and $\|F\| \leqslant \|h_\alpha\|_{\tilde{W}[0,1]}$. In fact equality holds here: this is because $W[0, 1]$ is a dense subspace of $\tilde{W}[0, 1]$.

7

Linear operators

After Newton's success in giving a mathematical account of planetary motion, scientists were emboldened to attempt the description of many physical phenomena using the differential calculus. The 'infinitesimal' viewpoint gave a fruitful way of formulating laws governing everyday processes or their idealizations. Wave motion, fluid flow and the conduction of heat were analysed. Their description is harder than that of celestial motions, if only because the world of our experience has such an abundance of detail in comparison with the emptiness of space. Still, there is a miraculous feature of the fabric of the universe to hearten the mathematician: a great diversity of physical processes is well described by second-order linear partial differential equations. Much of functional analysis stems from the study of such equations.

Imagine that you are a pioneer of mathematical physics faced with a new class of equations of the general type $Lf = g$. Here g is supposed to be a known function of space and time variables, L is a linear differential operator and f is a function which is unknown but required to satisfy some initial or boundary conditions. Your first concern would be to devise a way of finding f. Even if you work out a bag of tricks that does well in practice, think of the difficulty of proving that a particular technique will always succeed for the class of equations under study. Furthermore, it is very unlikely that you will obtain a general solution in a form explicit enough to tell you everything you want to know about the system. You will be faced with questions of a qualitative nature which the mathematics of the late eighteenth and early nineteenth centuries was scarcely in a position to formulate properly, never mind answer. Does every equation of your class *have* a solution? If so, is it unique? It is not enough to say that your equations describe a real physical process and therefore must admit a unique solution. Mathematical descriptions never correspond exactly to reality: they always involve idealization, and there is in addition the possibility that you have failed to take into account some essential feature – say, you have overlooked a boundary condition. If there is a unique solution, is it highly sensitive to small perturbations of the initial conditions? If it is, its usefulness is very doubtful since physical measurements are always approximate. Can one make qualitative deductions about the solution – for example, about its oscillatory nature – directly from the differential equations, thereby obviating the need to find explicit solutions, at least for some purposes?

In attacking these questions you might seek inspiration from linear algebra: $Lf = g$ does resemble $Ax = b$. However, the differences between finite and infinite dimensions are great, and in 1800 they were greater still. Nowadays we think of systems of linear equations in geometric terms: we speak of subspaces, linear spans, orthogonality and dimension. Then, habits of thought were quite different and were almost entirely based on the use of determinants. In infinite dimensions determinants can play only a very limited role (in this and most books on functional analysis, none at all). The change in outlook has come about through the stimulus of just such questions as the above. Modern linear algebra is pregnant with infinite dimensions.

Nineteenth-century analysts did make great progress with these questions. The striving to present their results with simplicity, generality and beauty gradually brought about the evolution of the concepts now classed as functional analysis. By the 1920s the essential components of the theory – measure theory, linear algebra and topology – were well enough understood for a synthesis to take place, and we now enjoy a theory whose elegance and completeness gives the appearance of finality. This is not to say that mathematical physics has solved all its problems. There are many areas of active research. However, the basic structures presented in this book have proved their worth and seem likely to remain fundamental to applied analysis for the near future.

In this chapter we shall see several types of linear operator in infinite dimensions, we shall consider some of the main similarities to and differences from finite matrices and establish some basic notions.

7.1 Definitions   If E, F are vector spaces over a field k, a *linear operator from E to F* is a mapping $T: E \to F$ such that

$$T(\lambda x + \mu y) = \lambda Tx + \mu Ty$$

for all $\lambda, \mu \in k$ and all $x, y \in E$.

A *linear operator on E* is a linear operator from E to E.

If E, F are normed spaces, a linear operator $T: E \to F$ is said to be *bounded* if there exists $M \geqslant 0$ such that

$$\|Tx\| \leqslant M \|x\|, \quad \text{all } x \in E.$$

For such a T the *norm*, or *operator norm* of T is the non-negative real number

$$\sup\{\|Tx\| : x \in E, \|x\| \leqslant 1\}$$

and is denoted by $\|T\|$.

The *kernel*, Ker T, of a linear operator $T: E \to F$ is the subspace $\{x \in E: Tx = 0\}$ of E, and the *range* of T is the subspace $\{Tx: x \in E\}$ of F. □

$\|T\|$ can be thought of as the largest factor by which T stretches any vector, though strictly it is a sup, not a max. Note that, for any $x \in E$,

$$\|Tx\| \leqslant \|T\| \|x\|.$$

In the case that $x \neq 0$ this follows from the fact that $\|Ty\| \leqslant \|T\|$ for the unit vector $y = x/\|x\|$.

7.2 *Examples*

(i) *A multiplication operator.* Define M on $L^2(a, b)$ by

$$(Mx)(t) = f(t)x(t)$$

where $f \in C[a, b]$. M is clearly linear, and, for any x,

$$\|Mx\|^2 = \int_a^b |f(t)|^2 |x(t)|^2 \, dt$$

$$\leqslant \sup_{a \leqslant t \leqslant b} |f(t)|^2 \int_a^b |x(t)|^2 \, dt$$

$$= \|f\|_\infty^2 \|x\|^2.$$

Thus M is bounded and $\|M\| \leqslant \|f\|_\infty$. In fact $\|M\| = \|f\|_\infty$. To prove the other inequality take x to be a scalar multiple of the characteristic function of a very small interval about a point for which $|f(t)|$ attains its maximum modulus.

(ii) *An integral operator.* Let $a, b, c, d \in \mathbb{R}$, let

$$k: [c, d] \times [a, b] \to \mathbb{C}$$

be continuous, and define $K: L^2(a, b) \to L^2(c, d)$ by

$$(Kx)(t) = \int_a^b k(t, s)x(s) \, ds, \quad c < t < d.$$

K is clearly linear. By the Cauchy–Schwarz inequality, we have, for fixed $t \in (c, d)$,

$$|Kx(t)|^2 \leqslant \left\{ \int_a^b |k(t, s)|^2 \, ds \right\} \left\{ \int_a^b |x(s)|^2 \, ds \right\}.$$

Hence

$$\|Kx\|^2 \leqslant \left\{ \int_c^d \int_a^b |k(t,s)|^2 \, ds \, dt \right\} \|x\|^2.$$

Thus K is bounded and

$$\|K\| \leqslant \left\{ \int_c^d \int_a^b |k(t,s)|^2 \, ds \, dt \right\}^{1/2}.$$

(This time the inequality is usually strict.)

(iii) *A differential operator.* Let $\mathscr{D}$ be the space of differentiable functions $f \in L^2(-\infty, \infty)$ such that $f' \in L^2(-\infty, \infty)$. Then

$$\frac{d}{dx} : \mathscr{D} \to L^2(-\infty, \infty)$$

is a linear operator. It is not bounded with respect to the L^2 norm on $\mathscr{D}$: functions f_n with graphs of the form shown in Diagram 7.1 can be chosen so that $f_n \to 0$ in L^2 but $\|f_n'\| \to \infty$.

This example illustrates significant features of differential operators. Their domains of definition have to be defined with care, and they are unbounded with respect to L^2 type inner products. As many of the most important operators in physics are differential operators one might conclude that it would be a waste of effort to develop the theory of bounded operators. This would be a mistaken conclusion: a juster one would be that some ingenuity will be needed to make the general theory fit the desired application. There are three main devices for getting the theory of bounded operators to yield information about differential operators. One is the observation that, though differential operators are unbounded, their inverses are often bounded. Another way of putting this is to say that most differential equations can be rewritten as integral equations. A second way is to impose a 'stronger' norm (such as the one determined by the inner product $\int f\bar{g} + f'\bar{g}'$ in Problem 1.2) on the domain space in order to force

Diagram 7.1

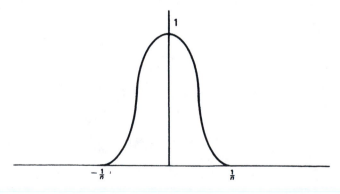

the operator to be bounded. The third and most radical approach is to develop the theory of operators which, though possibly unbounded, are *closed*, in the sense of having a graph which is a closed subset of the Cartesian product of the domain and codomain. Most differential operators have this property.

(iv) *A shift operator.* Define S on ℓ^2 by

$$S(x_1, x_2, x_3, \ldots) = (0, x_1, x_2, \ldots).$$

S is clearly an isometry: $\|Sx\| = \|x\|$ for all $x \in \ell^2$. Thus S is bounded and $\|S\| = 1$.

There is also the *backward shift operator* S^* on ℓ^2, defined by

$$S^*(x_1, x_2, x_3, \ldots) = (x_2, x_3, x_4, \ldots).$$

S^* is bounded and $\|S^*\| = 1$, but S^* is not an isometry. $\qquad \square$

7.3 *Exercise* Let E be a finite-dimensional inner product space, having orthonormal basis $e_1, \ldots, e_n$, and let F be any normed space. Show that any linear operator $A: E \to F$ is bounded and satisfies

$$\|A\| \leqslant \left\{ \sum_{j=1}^{n} \|Ae_j\|^2 \right\}^{1/2}. \qquad \square$$

Continuity is the same as boundedness for linear operators on normed spaces.

7.4 *Theorem* Let E, F be normed spaces and let $T: E \to F$ be a linear operator. The following are equivalent:

(i) T is continuous;

(ii) T is continuous at 0;

(iii) T is bounded. $\qquad \square$

To prove this theorem one need only replace modulus signs by norms in the proof of Theorem 6.3.

7.1 The Banach space $\mathcal{L}(E, F)$

Many properties of linear functionals generalize easily to linear operators. The analogue of the dual space is the space of all continuous linear operators from one normed space into another.

7.5 *Theorem* If E, F are normed spaces then the space $\mathcal{L}(E, F)$ of all continuous linear operators from E to F is itself a normed space with respect to pointwise operations and the operator norm. If, further, F is a Banach space then so is $\mathcal{L}(E, F)$. $\qquad \square$

We shall denote the composition of operators $A: E \to F$, $B: F \to G$ by BA. Thus, if $x \in E$,

$$BAx = B(Ax) \in G.$$

7.6 Theorem Let E, F and G be normed spaces. If $A \in \mathscr{L}(E, F)$ and $B \in \mathscr{L}(F, G)$ then $BA \in \mathscr{L}(E, G)$ and

$$\|BA\| \leqslant \|B\| \|A\|.$$

Proof. BA is clearly linear and (being a composition of continuous mappings) continuous. For any $x \in E$,

$$\|BAx\|_G = \|B(Ax)\|_G$$
$$\leqslant \|B\| \|Ax\|_F$$
$$\leqslant \|B\| \|A\| \|x\|_E.$$

Hence

$$\|BA\| \leqslant \|B\| \|A\|. \qquad \square$$

We shall abbreviate $\mathscr{L}(E, E)$ to $\mathscr{L}(E)$. For $A \in \mathscr{L}(E)$ we denote $AA, AAA, \ldots$ by $A^2, A^3, \ldots$ etc.

7.7 Exercise Let E be a normed space and let $A \in \mathscr{L}(E)$. Show that, for any $n \in \mathbb{N}$,

$$\|A^n\| \leqslant \|A\|^n. \qquad \square$$

7.2   Inverses of operators

Solving equations is the same as inverting mappings. The linear equation $Ax = y$ has the unique solution $x = A^{-1}y$. In finite dimensions there is a comprehensive theory, based on determinants, which tells us everything we could want to know about the existence of A^{-1} and its evaluation. In infinite dimensions things are more complicated, but the notion of an inverse operator is still a helpful one.

We shall denote the identity operator on a space E by I_E, or simply I.

7.8 Definition Let E, F be normed linear spaces. An operator $A \in \mathscr{L}(E, F)$ is *invertible* if there exists $B \in \mathscr{L}(F, E)$ such that

$$AB = I_F \quad \text{and} \quad BA = I_E.$$

Such a B is unique, when it exists: it is called the *inverse* of A and is denoted by A^{-1}. $\qquad \square$

Note that A^{-1}, when it exists, is the ordinary (set-theoretic) inverse mapping of A. It may, however, happen that an operator is bijective, so that it possesses an inverse mapping (which is necessarily linear), and yet this inverse mapping is not bounded. A is then not invertible according to our definition. See Problem 7.16.

If A is an operator on a finite-dimensional normed space E the following are equivalent.

   (i) A is invertible;
   (ii) A is injective;
   (iii) A is surjective;
   (iv) there exists $B \in \mathscr{L}(E)$ such that $AB = I$;
   (v) there exists $B \in \mathscr{L}(E)$ such that $BA = I$.

In infinite dimensions these statements are all inequivalent.

7.9 Examples   (i) The shift operators S and S^* on ℓ^2 defined in Example 7.2(iv) satisfy

$$S^*S = I, \qquad SS^* \neq I.$$

It follows that neither S nor S^* is invertible, though S has a left inverse (and is injective) while S^* has a right inverse (and is surjective).

  (ii) The multiplication operator M on $L^2(0, 1)$ defined by

$$(Mx)(t) = tx(t), \qquad 0 < t < 1,$$

is injective but not surjective, and hence not invertible. For suppose $Mx = 0$: then $tx(t) = 0$ for almost all $t \in (0, 1)$, and hence $x(t) = 0$ for almost all $t \in (0, 1)$, so that x is the [equivalence class of the] zero function. However, there is no $x \in L^2(0, 1)$ such that Mx is the function which is identically 1, since the function $1/t$ is not square-integrable over $(0, 1)$. $\square$

  Invertibility has clearly become a more delicate matter than it was for finite matrices. Even so, there is one simple but surprisingly useful way of producing inverses. We generalize the fact that

$$(1 - z)^{-1} = \sum_{n=0}^{\infty} z^n$$

for any $z \in \mathbb{C}$ such that $|z| < 1$.

7.10 Theorem   Let E be a Banach space and let $A \in \mathscr{L}(E)$. If $\|A\| < 1$ then $I - A$ is invertible and

$$(I - A)^{-1} = \sum_{n=0}^{\infty} A^n \tag{7.1}$$

in the normed space $\mathscr{L}(E)$.

  Note that A^0 is defined to be the identity operator, whatever the operator A. Thus (7.1) can be written

$$(I - A)^{-1} = \lim_{n \to \infty} (I + A + A^2 + \cdots + A^n)$$

with respect to the operator norm on $\mathscr{L}(E)$.

Proof. For any $x \in E$, the sequence $((I + A + \cdots + A^n)x)_{n=1}^{\infty}$ is Cauchy in E, for if $m > n$,

$$\|(I + A + \cdots + A^m)x - (I + A + \cdots + A^n)x\|$$
$$= \|A^{n+1}x + \cdots + A^m x\|$$
$$\leqslant \|A^{n+1}x\| + \cdots + \|A^m x\|$$
$$\leqslant \|A^{n+1}\|\|x\| + \cdots + \|A^m\|\|x\|$$
$$\leqslant \|x\|\{\|A\|^{n+1} + \cdots + \|A\|^m\}$$
$$\leqslant \|x\| \sum_{r=n+1}^{\infty} \|A\|^r$$
$$= \|x\| \frac{\|A\|^{n+1}}{1 - \|A\|}. \tag{7.2}$$

Since $\|A\|^{n+1} \to 0$ as $n \to \infty$ it follows that the left hand side tends to zero as $m, n \to \infty$ with $m > n$.

Since E is a Banach space, the Cauchy sequence $((I + A + \cdots + A^n)x)_n$ converges to a limit $Tx \in E$. T is clearly a linear operator on E. On letting $m \to \infty$ in the inequality (7.2) we find

$$\|Tx - (I + A + \cdots + A^n)x\| \leqslant \|x\| \frac{\|A\|^{n+1}}{1 - \|A\|}.$$

This shows that $T - (I + A + \cdots + A^n)$ is a bounded linear operator, and hence that T is. Furthermore

$$\|T - (I + A + \cdots + A^n)\| \leqslant \frac{\|A\|^{n+1}}{1 - \|A\|},$$

which implies that

$$I + A + \cdots + A^n \to T$$

as $n \to \infty$.

It remains to show that $T = (I - A)^{-1}$. For any $x \in E$, continuity of $I - A$ gives

$$(I - A)Tx = (I - A) \lim_{n \to \infty} (I + A + \cdots + A^n)x$$

$$= \lim_{n \to \infty} (I - A)(I + A + \cdots + A^n)x$$

$$= \lim_{n \to \infty} x - A^{n+1}x.$$

We have

$$\|A^{n+1}x\| \leqslant \|A^{n+1}\|\|x\|$$
$$\leqslant \|A\|^{n+1}\|x\|$$

and so $A^{n+1}x \to 0$ as $n \to \infty$. Thus $(I-A)Tx = x$. A similar calculation shows that $T(I-A)x = x$, and so $T = (I-A)^{-1}$. □

7.11 Corollary   Let E be a Banach space. The set of invertible operators on E is open in $\mathscr{L}(E)$.

Proof. Observe that if X and Y are invertible in $\mathscr{L}(E)$ then so is XY (since $Y^{-1}X^{-1}$ is an inverse). Suppose A is invertible and let B be any element of the open ball in $\mathscr{L}(E)$ with centre A and radius $\|A^{-1}\|^{-1}$: thus

$$\|B - A\| < \|A^{-1}\|^{-1}.$$

Then

$$\|(B-A)A^{-1}\| \leqslant \|B-A\|\|A^{-1}\| < 1,$$

so that, by Theorem 7.10, the operator

$$I + (B-A)A^{-1} = BA^{-1}$$

is invertible. Thus

$$B = (BA^{-1})A$$

is invertible. The set of invertible elements in $\mathscr{L}(E)$ thus contains a ball centred at A, and so is open. □

7.12 Exercise   Let K be the integral operator on $L^2(0, 1)$ defined by

$$Kf(t) = \int_0^t (t-s)f(s)\,ds, \qquad 0 < t < 1.$$

Show that $\|K\| < 1$ and that

$$K^n f(t) = \int_0^t \frac{(t-s)^{2n-1}}{(2n-1)!}\, f(s)\,ds.$$

Hence solve for $f \in L^2(0, 1)$ the integral equation

$$f(t) = g(t) + \int_0^t (t-s)f(s)\,ds$$

where g is a given function in $L^2(0, 1)$.

Observe that the integral equation can be written $(I-K)f = g$. Use the geometric series to obtain an expression for

$$f = (I-K)^{-1}g.$$ □

7.3   Adjoint operators

The application of duality to the study of an $m \times n$ matrix A forces one to introduce the *conjugate transpose* A^* of A, in view of the relation

$$(Ax, y) = (x, A^*y)$$

valid for all $x \in \mathbb{C}^n$, $y \in \mathbb{C}^m$. The Riesz–Fréchet theorem enables us to define a counterpart in any Hilbert space.

7.13 Theorem Let $A \in \mathcal{L}(E, F)$ where E, F are Hilbert spaces. There exists a unique operator $A^* \in \mathcal{L}(F, E)$ such that

$$(Ax, y)_F = (x, A^*y)_E \tag{7.3}$$

for all $x \in E$, $y \in F$.

Proof. Consider any $y \in F$. The mapping

$$x \mapsto (Ax, y)_F$$

is a continuous linear functional on E, and so, by the Riesz–Fréchet theorem, there is a unique element $z \in E$ such that $(Ax, y)_F = (x, z)_E$ for all $x \in E$. Define A^*y to be z. Then A^* is a mapping from F to E and (7.3) is satisfied. Let us show that A^* is linear. For any $y, z \in F$ and $\lambda, \mu \in \mathbb{C}$ we have, for any $x \in E$,

$$
\begin{aligned}
(x, A^*(\lambda y + \mu z))_E &= (Ax, \lambda y + \mu z)_F \\
&= \bar{\lambda}(Ax, y)_F + \bar{\mu}(Ax, z)_F \\
&= \bar{\lambda}(x, A^*y)_E + \bar{\mu}(x, A^*z)_E \\
&= (x, \lambda A^*y + \mu A^*z)_E.
\end{aligned}
$$

Hence, by Theorem 1.5(iv),

$$A^*(\lambda y + \mu z) = \lambda A^*y + \mu A^*z.$$

That is, A^* is linear.

To see that A^* is bounded observe that, for any $y \in F$,

$$
\begin{aligned}
\|A^*y\|^2 &= (A^*y, A^*y) \\
&= (AA^*y, y) \\
&\leqslant \|AA^*y\| \|y\|,
\end{aligned}
$$

by the Cauchy–Schwarz inequality. Since $\|AA^*y\| \leqslant \|A\| \|A^*y\|$, this implies that

$$\|A^*y\|^2 \leqslant \|A\| \|A^*y\| \|y\|.$$

If $\|A^*y\| > 0$ we may divide through to obtain

$$\|A^*y\| \leqslant \|A\| \|y\|,$$

and since this inequality is trivially true if $\|A^*y\| = 0$, it holds always. It follows that A^* is a bounded linear operator, and

$$\|A^*\| \leqslant \|A\|. \tag{7.4}$$

Lastly, we must show that A^* is uniquely determined by the relation (7.3). That is, if (7.3) is satisfied when $A^* = B_1$ and when $A^* = B_2$, where $B_1, B_2 \in \mathcal{L}(E, F)$, then it must be the case that $B_1 = B_2$. This is indeed so, for the hypothesis implies that $(x, B_1 y) = (x, B_2 y)$ for any $x \in E$, $y \in F$, and we may apply Theorem 1.5(iv) to conclude that $B_1 y = B_2 y$. □

A^* is called the *adjoint* of A. Let us calculate some adjoints.

7.14 Examples (i) The multiplication operator M on $L^2(a, b)$ defined by

$$Mx(t) = f(t)x(t), \qquad -\infty < a < t < b < \infty,$$

where $f \in C[a, b]$, has adjoint M^* which is also a multiplication operator. Indeed, the defining relation for M^* is

$$(x, M^*y) = (Mx, y),$$

that is,

$$\int_a^b x(t)(M^*y(t))^- \, dt = \int_a^b f(t)x(t)y(t)^- \, dt$$

for all $x, y \in L^2(a, b)$. This shows that

$$(M^*y(t))^- = f(t)y(t)^-$$

almost everywhere, and hence

$$M^*y(t) = f(t)^- y(t).$$

Thus M^* is the operation of multiplication by the complex conjugate of f. In particular, if f is real-valued, $M^* = M$.

(ii) The adjoint of an integral operator is an integral operator. Let $K : L^2(a, b) \to L^2(c, d)$ be the integral operator with kernel k, as in Example 7.2(ii). The defining relation for K^* is

$$(x, K^*y) = (Kx, y),$$

that is,

$$\int_a^b x(s)(K^*y(s))^- \, ds = \int_c^d Kx(t)y(t)^- \, dt$$

$$= \int_c^d \int_a^b k(t, s)x(s)y(t)^- \, ds \, dt$$

$$= \int_a^b \int_c^d x(s)k(t, s)y(t)^- \, dt \, ds.$$

The reversal of the order of integration is justified by Fubini's theorem (de Barra, 1981). As this holds for all $x \in L^2(a, b)$ and $y \in L^2(c, d)$ we must have

$$(K^*y(s))^- = \int_c^d k(t, s)y(t)^- \, dt,$$

or, interchanging the role of s and t,

$$K^*y(t) = \int_c^d k(s, t)^- y(s) \, ds$$

for almost all t. Thus K^* is the integral operator with kernel k^*, where $k^*(t, s) = k(s, t)^-$.

(iii) The adjoint of the shift operator S on ℓ^2 (Example 7.2(iv)) is the backward shift operator S^*.

We leave this as an instructive exercise.   □

7.15 Theorem $A^{**} = A$ and $\|A^*\| = \|A\|$ for any pair E, F of Hilbert spaces and any $A \in \mathscr{L}(E, F)$.

Proof. A^{**} of course means B^* where $B = A^*$. With this notation, $B \in \mathscr{L}(F, E)$ and $B^* \in \mathscr{L}(E, F)$. On applying the definition (7.3) with A replaced by B we have, for $x \in F$, $y \in E$,

$$(x, B^*y)_F = (Bx, y)_E$$
$$= (A^*x, y)_E$$
$$= (y, A^*x)_E^-$$
$$= (Ay, x)_F^-$$
$$= (x, Ay)_F$$

and hence

$$A^{**}y = B^*y = Ay,$$

so that $A^{**} = A$.

From (7.4) we have

$$\|A^*\| \leqslant \|A\|.$$

Since this applies to any bounded linear operator we may replace A by A^* to obtain

$$\|A^{**}\| \leqslant \|A^*\|,$$

and in view of the first part of the theorem, this means that

$$\|A\| \leqslant \|A^*\|.$$

Hence

$$\|A\| = \|A^*\|.$$ □

7.16 Exercise   Let E, F, G be Hilbert spaces. Show that

(i) if $A \in \mathscr{L}(E, F)$, $B \in \mathscr{L}(F, G)$ then

$$(BA)^* = A^*B^*,$$

(ii) if $A, B \in \mathscr{L}(E, F)$ and $\lambda, \mu \in \mathbb{C}$ then

$$(\lambda A + \mu B)^* = \bar{\lambda}A^* + \bar{\mu}B^*.$$ □

7.4   Hermitian operators

Just as in finite dimensions, operators which coincide with their own adjoints are particularly tractable: there are 'diagonalization' results for them.

7.17 Definition   Let H be a Hilbert space and let $A \in \mathcal{L}(H)$. A is *Hermitian* or *self-adjoint* if $A = A^*$.   □

From Examples 7.14 we can readily produce some infinite-dimensional Hermitian operators. Multiplication by a bounded real-valued continuous function on (a, b) is a Hermitian operator on $L^2(a, b)$. The integral operator with kernel function

$$k : (a, b) \times (a, b) \to \mathbb{C}$$

is a Hermitian operator on $L^2(a, b)$ provided that it is bounded and k satisfies

$$k(t, s) = \overline{k(s, t)}, \qquad a < s, t < b.$$

This type of Hermitian operator will play a central role in our study of certain boundary value problems in Chapters 10 and 11.

For the case of a Hermitian matrix, the result which follows is obvious from diagonalization. Here it will be a step in the proof of a diagonalization result for certain Hermitian operators.

7.18 Theorem   If A is a Hermitian operator on a Hilbert space H then

$$\|A\| = \sup_{\|x\| = 1} |(Ax, x)|. \tag{7.5}$$

Proof. For any $x \in H$ such that $\|x\| = 1$,

$$|(Ax, x)| \leqslant \|Ax\| \|x\|$$
$$\leqslant \|A\| \|x\|^2$$
$$= \|A\|.$$

Hence

$$\|A\| \geqslant \sup_{\|x\| = 1} |(Ax, x)|.$$

Let m denote the right hand side of (7.5), so that

$$|(Au, u)| \leqslant m\|u\|^2$$

for any $u \in H$. Then for any x, y,

$$(A(x \pm y), x \pm y) = (Ax, x) \pm 2 \operatorname{Re}(Ax, y) + (Ay, y).$$

Thus

$$4 \operatorname{Re}(Ax, y) = (A(x + y), x + y) - (A(x - y), x - y)$$
$$\leqslant m(\|x + y\|^2 + \|x - y\|^2)$$
$$= 2m(\|x\|^2 + \|y\|^2),$$

by the parallelogram law. Replace x by λx where $|\lambda| = 1$ and λ is chosen so that $\lambda(Ax, y) \geqslant 0$. We obtain

$$|(Ax, y)| \leqslant \tfrac{1}{2}m(\|x\|^2 + \|y\|^2).$$

Suppose that $Ax \neq 0$. Since the above holds for all $x, y \in H$ we may choose y to be given by

$$y = \frac{\|x\|}{\|Ax\|} Ax.$$

Since $\|y\| = \|x\|$ the inequality yields

$$\|x\| \|Ax\| \leqslant m \|x\|^2,$$

and hence (since $x \neq 0$)

$$\|Ax\| \leqslant m \|x\|.$$

This holds trivially when $Ax = 0$, hence is true for all x. From the definition of operator norm

$$\|A\| \leqslant m$$
$$= \sup_{\|x\| = 1} |(Ax, x)|. \qquad \square$$

7.19 Exercise   Show that (7.5) need not hold for non-Hermitian A by considering A on $\mathbb{C}^2$ given by $A(x_1, x_2) = (0, x_1)$. $\qquad \square$

7.5   The spectrum

If A is an $n \times n$ complex matrix then A has at least one eigenvalue: that is, there exists $\lambda \in \mathbb{C}$ such that $Ax = \lambda x$ for some non-zero vector x. The *spectrum* of A consists of the set of all eigenvalues of A and is an important tool, since all diagonalization results for matrices are based on the use of eigenvalues. It is clearly a high priority in extending matrix theory to infinite dimensions to establish the right analogue of the set of eigenvalues.

The first thing to notice is that bounded linear operators need not have *any* eigenvalues. Recall the shift operator S on ℓ^2 from Example 7.2(iv):

$$S(x_1, x_2, \ldots) = (0, x_1, \ldots).$$

It is readily seen that $Sx = \lambda x$ implies $x = 0$, and hence S has no eigenvalues. Nor has the multiplication operator M, Example 7.2(i). The collection of eigenvalues of an operator in infinite dimensions is not such a useful notion in general: a better one is the following.

7.20 Definition   Let $A \in \mathcal{L}(E)$ where E is a Banach space. The *spectrum* $\sigma(A)$ of A is the set of $\lambda \in \mathbb{C}$ such that $\lambda I - A$ is not invertible. $\qquad \square$

The spectrum of any operator A on a Banach space $E \neq \{0\}$ is non-empty: otherwise the mapping $\lambda \mapsto (\lambda I - A)^{-1}$ would be a bounded non-constant $\mathcal{L}(E)$-valued entire function on $\mathbb{C}$, contrary to a version of Liouville's theorem which holds for Banach-space-valued functions. If E is finite dimensional then $\sigma(A)$ coincides with the set of eigenvalues of A since $\lambda I - A$ is invertible if and only if $\text{Ker}(\lambda I - A) = \{0\}$.

We shall continue to speak of eigenvalues of A in the sense implicit above: λ is an eigenvalue of A if $Ax = \lambda x$ for some $x \neq 0$. Every eigenvalue λ of A belongs to $\sigma(A)$, since $\lambda I - A$ has non-zero kernel and so cannot be invertible. The remarks above and the example below show that $\sigma(A)$ can contain non-eigenvalues.

7.21 Example Let $f : [a, b] \rightarrow \mathbb{C}$ be continuous, where $a < b \in \mathbb{R}$. The multiplication operator M_f on $L^2(a, b)$, defined by

$$(M_f x)(t) = f(t)x(t), \qquad a < t < b,$$

is a bounded operator. $\sigma(M_f)$ consists of the image of f. If $\lambda \notin f([a, b])$ then $\lambda I - M_f$ has bounded inverse $M_{(\lambda - f)^{-1}}$, and so $\lambda \notin \sigma(M_f)$. On the other hand, if $\lambda = f(t_0)$, some $t_0 \in [a, b]$, then $\lambda \in \sigma(M_f)$. Otherwise $\lambda I - M_f$ has a bounded inverse T. Pick an interval J_n about t_0 in $[a, b]$, of length $\delta_n > 0$, such that $|f(t) - \lambda| < 1/n$ for $t \in J_n$, and define

$$g_n(t) = \begin{cases} \delta_n^{-1/2} & \text{if } t \in J_n \\ 0 & \text{otherwise.} \end{cases}$$

Then $(\lambda I - M_f)g_n \rightarrow 0$ as $n \rightarrow \infty$, but $T(\lambda I - M_f)g_n = g_n$, which has norm 1 for all n, contradicting continuity of T.

This shows that $\sigma(M_f) = f([a, b])$. However for many f (in particular, for $f(t) = t$), M_f has no eigenvalues. $\qquad \square$

7.22 Theorem Let E be a Banach space and let $A \in \mathscr{L}(E)$. Then $\sigma(A)$ is a compact subset of $\mathbb{C}$ and is contained in the closed disc of centre 0, radius $\|A\|$.

Proof. Define a mapping $F : \mathbb{C} \rightarrow \mathscr{L}(E)$ by $F(\lambda) = \lambda I - A$. We have

$$\|F(\lambda) - F(\mu)\| = |\lambda - \mu|,$$

and hence F is continuous with respect to the operator norm on $\mathscr{L}(E)$. By definition,

$$\sigma(A) = F^{-1}(\mathscr{L}(E) \backslash G),$$

where G is the group of invertible elements in $\mathscr{L}(E)$. By Corollary 7.11, G is open and hence $\mathscr{L}(E) \backslash G$ is closed in $\mathscr{L}(E)$. Hence $\sigma(A)$ is closed in $\mathbb{C}$.

Suppose $|\lambda| > \|A\|$. Then $\|\lambda^{-1}A\| < 1$, and so, by Theorem 7.10, $I - \lambda^{-1}A$ is invertible, and hence $\lambda I - A$ is invertible: that is, $\lambda \notin \sigma(A)$. Thus $\sigma(A)$ is contained in the closed disc of centre 0, radius $\|A\|$. Since $\sigma(A)$ is closed and bounded in $\mathbb{C}$ it is compact. $\qquad \square$

7.6   Infinite matrices

If functional analysis can be regarded as a generalization of linear algebra to infinite dimensions, the reader may wonder why infinite matrices have not yet put in an appearance. Historically they were much studied, but as experience of the subject was gained mathematicians realized that in developing the general theory it was better to avoid the use of co-ordinates, and hence of matrices, wherever possible. Nevertheless for certain classes of operators the use of matrices can be an aid to the intuition, particularly when the matrices exhibit some sort of pattern.

7.23 Definition   Let E, F be Hilbert spaces and let $A \in \mathscr{L}(E, F)$. The *matrix of A with respect to orthonormal bases* $(e_n)_{n \in \mathbb{N}}$, $(f_m)_{m \in \mathbb{N}}$ of E, F respectively is the array $[a_{ij}]_{i,j=1}^{\infty}$ of complex numbers given by

$$a_{ij} = (Ae_j, f_i). \qquad \square$$

If we think of i, j as row and column indices in the usual way then the jth column of $[a_{ij}]$ consists of the components with respect to the basis (f_m) of Ae_j.

One disadvantage of a matrix representation of an operator is that it can be hard to recognize whether an infinite matrix does represent a bounded operator. It is necessary that each column be an ℓ^2 sequence, and that the ℓ^2-norms of the columns be uniformly bounded, but this is not sufficient (see Problem 7.42). Against this, if an operator has, say, a diagonal matrix with respect to a suitable orthonormal basis, it can be a distinct help to visualize it in this way.

7.24 Example   Let $k: \mathbb{R} \to \mathbb{C}$ be continuous and 2π-periodic and let the integral operator K on $L^2(-\pi, \pi)$ be given by

$$Kx(t) = \int_{-\pi}^{\pi} k(t - s)x(s)\, ds.$$

Then the matrix of K with respect to the orthonormal basis $((2\pi)^{-1/2} e^{int})_{n \in \mathbb{Z}}$ is the (doubly infinite) diagonal matrix with the Fourier coefficients of k as diagonal entries. Indeed, if e_n denotes the basis element with index n,

$$Ke_n(t) = \int_{-\pi}^{\pi} k(t - s)e_n(s)\, ds$$

$$= \frac{1}{\sqrt{(2\pi)}} \int_{-\pi}^{\pi} k(t - s)\, e^{ins}\, ds$$

$$= \frac{1}{\sqrt{(2\pi)}} \int_{-\pi}^{\pi} k(\tau)\, e^{in(t - \tau)}\, d\tau$$

$$= \frac{e^{int}}{\sqrt{(2\pi)}} \int_{-\pi}^{\pi} k(\tau) \, e^{-int} \, d\tau$$

$$= c_n e_n(t)$$

where c_n is the nth Fourier coefficient of k.   □

In many books on functional analysis infinite matrices are barely mentioned, but we shall have a significant use for them in Chapter 15.

7.7   Problems

Bounded operators

7.1. Let $(e_n)_1^\infty$ be a complete orthonormal sequence in a Hilbert space H and let $\lambda_n \in \mathbb{C}$ for $n \in \mathbb{N}$. Show that there is a bounded linear operator D on H such that $De_n = \lambda_n e_n$, all $n \in \mathbb{N}$, if and only if (λ_n) is a bounded sequence. What is $\|D\|$, when defined?

7.2. Let E be a Banach space and let $A, B \in \mathcal{L}(E)$. Define $T: \mathcal{L}(E) \to \mathcal{L}(E)$ by

$$TX = AXB.$$

Show that T is a linear operator and is bounded with respect to the operator norm on $\mathcal{L}(E)$, with $\|T\| \le \|A\| \|B\|$.

7.3. Let X, Y be compact Hausdorff spaces and $\alpha: X \to Y$ be a continuous mapping. Let $C(X)$ be the Banach space of continuous $\mathbb{C}$-valued functions on X with supremum norm and let $T: C(Y) \to C(X)$ be the operation of composition with α – that is,

$$(Tf)(x) = f \circ \alpha(x) = f(\alpha(x)),$$

all $f \in C(Y), x \in X$. Show that T is a bounded linear operator and $\|T\| \le 1$.

7.4. Show that the range of a bounded linear operator need not be closed by considering the multiplication operator

$$(Mx)(t) = tx(t)$$

on $L^2(0, 1)$ (show that the range of M is dense in $L^2(0, 1)$ but does not contain the constant function 1).

7.5. Is d/dz a bounded linear operator on RH^2?

7.6. The operator S^* on RH^2 is defined by

$$(S^*f)(z) = \frac{1}{z}\{f(z) - f(0)\},$$

with the removable singularity of the rational function on the right hand side removed. Prove that

$$(S^*f, g) = (f, zg(z))$$

in RH^2 (use the fact that the function g^* defined by $g^*(z) = \overline{g(\bar{z})}$ is analytic for $|z| \le 1$). Deduce that $\|S^*\| \le 1$.

7.7. Let $\mathscr{D}$ be as in Example 7.2(iii). Show that the linear operator

$$\frac{d}{dx} : \mathscr{D} \to L^2(-\infty, \infty)$$

is unbounded with respect to the L^2-norm on $\mathscr{D}$ but is bounded with respect to the inner product

$$(f, g) = \int_{-\infty}^{\infty} f(t)\overline{g(t)} + f'(t)\overline{g'(t)} \, dt$$

on $\mathscr{D}$.

7.8. The *Volterra operator* V on $L^2(0, 1)$ is the operator of indefinite integration:

$$(Vx)(t) = \int_0^t x(s) \, ds, \qquad 0 < t < 1.$$

Show that V is a bounded linear operator and that $\|V\| \leqslant 1/\sqrt{2}$.

7.9. An operator is said to have *rank 1* if its range is one-dimensional. Let T be a bounded operator of rank 1 on a Hilbert space H, and let ψ be a non-zero vector in the range of T. Show that there exists $\varphi \in H$ such that

$$Tx = (x, \varphi)\psi, \qquad \text{all } x \in H$$

and that

$$\|T\| = \|\varphi\| \|\psi\|.$$

7.10. An operator K on $L^2(0, 1)$ is defined by

$$(Kx)(t) = \int_0^1 e^{t-s} x(s) \, ds, \qquad 0 < t < 1.$$

Use the result of Problem 7.9 to show that

$$\|K\| = \sinh 1.$$

7.11. Define an operator U on the inner product space RH^2 by

$$Uf(z) = \frac{(1 - |\alpha|^2)^{1/2}}{1 - \bar{\alpha}z} f\left(\frac{z - \alpha}{1 - \bar{\alpha}z}\right)$$

for all $z \in \mathbb{D}$, where $\alpha \in \mathbb{D}$ is fixed. Show that U is an isometry: that is,

$$\|Uf\| = \|f\|, \qquad \text{all } f \in RH^2.$$

Inverses

7.12. Show that the operator D of Problem 7.1, assumed bounded, is invertible if and only if

$$\inf_n |\lambda_n| > 0.$$

What is $\|D^{-1}\|$, when applicable?

7.13. Let $S*$ be the operator on RH^2 defined in Problem 7.6. Show that if $\alpha \in \mathbb{D}$ then $(I - \alpha S*)^{-1}$ exists and is given by $(I - \alpha S*)^{-1}f = g$, where

$$g(z) = \frac{zf(z) - \alpha f(\alpha)}{z - \alpha},$$

with the removable singularity on the right hand side removed.

7.14. Show that the rank 1 operator T of Problem 7.9 satisfies $T^2 = (\psi, \varphi)T$. Hence show that, if $(\psi, \varphi) \neq 1$, $I - T$ is invertible and find $(I - T)^{-1}$.

7.15. Let V be the Volterra operator on $L^2(0, 1)$ (see Problem 7.8). Prove by induction that

$$(V^n x)(t) = \int_0^t \frac{(t - s)^{n-1}}{(n - 1)!} x(s)\, ds.$$

Hence solve for f the integral equation

$$f(t) = \sin t + \int_0^t f(s)\, ds$$

by the method of Exercise 7.12.

7.16. Give an example of normed spaces E and F and a bijective operator $T \in \mathcal{L}(E, F)$ which is not invertible in the sense of Definition 7.8 (that is, T^{-1} is an unbounded operator).

[Take $E = C[0, 1]$, $T = V | E$, or alternatively use D of Problem 7.1 for a suitable choice of (λ_n), with $F = $ range D.]

7.17. If $A, B \in \mathcal{L}(E)$ and AB is invertible, does it follow that A and B are invertible?

7.18. Show that A and B are invertible in $\mathcal{L}(E)$ if and only if both AB and BA are invertible.

7.19. Show that if $A, B \in \mathcal{L}(E)$ and $I - AB$ is invertible then $I - BA$ is also invertible, with inverse $I + B(I - AB)^{-1}A$.

Adjoints

7.20. Find the adjoint of the operator D of Problem 7.1 when (λ_n) is a bounded sequence.

7.21. Let H, K be Hilbert spaces and let $T \in \mathcal{L}(H, K)$. Show that $T*T$ is Hermitian and that

$$\|T*T\| = \|T\|^2$$

(use Theorem 7.18).

7.22. Let A be an $n \times n$ Hermitian matrix acting on $\mathbb{C}^n$ with the standard inner product. By diagonalization or otherwise, show that $\|A\|$ is the maximum of the moduli of the eigenvalues of A. Hence find the norm of

$$\begin{bmatrix} 1 & 1 \\ 1 & 2 \end{bmatrix}.$$

7.23. Find the norm of $\begin{bmatrix} 1 & 1 \\ 0 & 1 \end{bmatrix}$, regarded as an operator on $\mathbb{C}^2$ with the standard inner product.

7.24. An operator T is defined on the Hilbert space $\mathbb{C}^{m \times n}$ of Exercise 1.4 by

$$TX = UXV, \quad \text{all } X \in \mathbb{C}^{m \times n},$$

where $U \in \mathbb{C}^{m \times m}$, $V \in \mathbb{C}^{n \times n}$ are fixed. Find the adjoint of T.

7.25. Prove that the adjoint of the shift operator S on ℓ^2 is the backward shift operator (cf. Example 7.2(iv)).

7.26. Let T be the rank 1 operator of Problem 7.9. What is T^*?

7.27. Let V be the Volterra operator (Problem 7.8). Show that $V + V^*$ has rank 1.

7.28. Let u be continuously differentiable on $[0, 1]$ and let M be the operation of multiplication by u on $W[0, 1]$:

$$Mf = uf, \quad f \in W[0, 1].$$

Show that, for twice continuously differentiable functions g, h on $[0, 1]$,

$$(Mf, g) = (f, h) \quad \text{for all } f \in W[0, 1]$$

if and only if

$$\begin{cases} h'' - h = \bar{u}(g'' - g) \\ h'(0) = \overline{u(0)}g'(0), \quad h'(1) = \overline{u(1)}g'(1). \end{cases}$$

By the method of variation of parameters, or otherwise, express M^* as an integral operator.

7.29. Let H be a Hilbert space and let $A \in \mathscr{L}(H)$. Prove that $(\text{Range } A)^{\perp} = \text{Ker } A^*$ and that $(\text{Ker } A)^{\perp}$ is the closure of Range A^*. Prove also that Ker $A^*A = \text{Ker } A$.

7.30. Let H be a Hilbert space and let $A \in \mathscr{L}(H)$. A sequence (x_n) in H satisfies $A^*x_n \to y$ in H. Prove that there is a sequence (y_n) in H such that $A^*Ay_n \to y$.

7.31. Prove that a bounded operator U on a Hilbert space is an isometry (i.e. preserves norms) if and only if $U^*U = I$.

7.32. Let φ be a unit vector in a Hilbert space H and let T be the rank 1 operator

$$Tx = (x, \varphi)\varphi, \quad x \in H.$$

Prove that $I - 2T$ is a Hermitian isometry.

7.33. Let x, y be vectors of the same norm in a Hilbert space H. Show that, if $(x, y) \in \mathbb{R}$, there is a Hermitian isometry U on H such that $Ux = y$.

The spectrum

7.34. Find the spectrum of the operator D of Problem 7.1, when (λ_n) is a bounded sequence.

7.35. Show that every $\lambda \in \mathbb{D}$, the open unit disc, is an eigenvalue of the backward shift operator S^* on ℓ^2 (see Example 7.2(iv)). Deduce that $\sigma(S^*) = \text{clos } \mathbb{D}$ (use Theorem 7.22).

7.36. Find the spectrum of the rank 1 operator T of Problem 7.9 (use Problem 7.14).

7.37. Let $A \in \mathscr{L}(H)$, H a Hilbert space. Show that

$$\sigma(A^*) = \{\bar{\lambda} : \lambda \in \sigma(A)\}.$$

7.38. Find $\sigma(S)$ where S is the shift operator on ℓ^2 (Example 7.2(iv)).

7.39. Let $A, B \in \mathscr{L}(E)$, E a Banach space. Show that $\sigma(AB)$ and $\sigma(BA)$ coincide, except possibly for the point 0: that is,

$$\sigma(AB) \backslash \{0\} = \sigma(BA) \backslash \{0\}.$$

Give an example in which $\sigma(AB) \neq \sigma(BA)$.

7.40. By considering the identity

$$(\lambda I - A)(\lambda I + A) = \lambda^2 I - A^2$$

for any $\lambda \in \mathbb{C}$ and $A \in \mathscr{L}(E)$, E a Banach space, show that $\sigma(A^2) = \{\lambda^2 : \lambda \in \sigma(A)\}$. Deduce that, for any $\lambda \in \sigma(A)$,

$$|\lambda| \leqslant \|A^2\|^{1/2}.$$

7.41. Can you replace 2 by 3 in the foregoing problem? By $n \in \mathbb{N}$? By -1?

Matrices

7.42. Let $A = [a_{ij}]_{i,j=1}^{\infty}$ where

$$a_{ij} = \begin{cases} 0 & \text{if } i \leqslant j, \\ \dfrac{1}{i-j} & i > j. \end{cases}$$

The ℓ^2 norms of the rows and columns of A are uniformly bounded (by $\pi/\sqrt{6}$). Does A define a bounded operator relative to an orthonormal basis of a Hilbert space? Try this problem again after reading Theorem 13.14 and the comments following it.

8

Compact operators

The study of the differential and integral equations which arise in applications of mathematics would be greatly advanced if we could say as much about general linear operators as we can about finite-dimensional ones. For example, the various canonical form theorems of linear algebra enable us to answer many questions with ease: are there analogues in infinite dimensions? The answer is a heavily qualified *yes*. Such is the diversity even of bounded linear operators on Hilbert space that we cannot expect anything of the generality and simplicity of, say, the Jordan form theorem. Nevertheless, there are classes of operators which are broad enough to include many operators of interest and still narrow enough to allow the proof of significant theorems. One such class consists of the compact operators. These include many integral operators and yet have properties close to those of matrices.

8.1 Definition   Let E, F be normed spaces and let $A: E \to F$ be linear. A is said to be *compact* if, for every bounded sequence $(x_n)_1^\infty$ in E, the sequence $(Ax_n)_1^\infty$ has a convergent subsequence in F.   □

A compact operator A is necessarily bounded. Otherwise there would exist a bounded sequence (x_n) in E such that $\|Ax_n\| \to \infty$, in which case (Ax_n) could have no convergent subsequence.

8.2 Examples   (i) Bounded finite rank operators are compact. The *rank* of an operator is defined to be the dimension of its range. Thus if $A \in \mathscr{L}(E, F)$ has finite rank then A maps E into some finite-dimensional subspace R of F. R is a normed space with respect to the restriction of the norm of F. By Corollary 2.14, closed bounded sets in R are compact. Thus, if (x_n) is a bounded sequence in E then $\{Ax_n : n \in \mathbb{N}\}$ is a bounded set in R, and hence its closure is compact. Thus $\{Ax_n : n \in \mathbb{N}\}$ has a cluster point y in R, and since we are dealing with metric spaces, it follows that some subsequence of (Ax_n) converges to y.

(ii) The identity operator I on an infinite-dimensional Hilbert space H is not compact. Choose (x_n) to be an infinite orthonormal sequence in H: then if $m \neq n$,

$$\|x_n - x_m\|^2 = \|x_n\|^2 - (x_n, x_m) - (x_m, x_n) + \|x_m\|^2$$
$$= 2.$$

Hence distinct terms of the sequence (x_n) are at a distance $\sqrt{2}$ from each other. The sequence (Ix_n) thus contains no Cauchy subsequence, and consequently no convergent subsequence.

(iii) Diagonal operators. Let $(e_n)_1^\infty$ be a complete orthonormal sequence in a Hilbert space H and let $(\lambda_n)_1^\infty$ be a bounded sequence of complex numbers. Let $A \in \mathscr{L}(H)$ be the operator whose matrix with respect to the basis (e_n) is $\mathrm{diag}\{\lambda_1, \lambda_2, \lambda_3, \ldots\}$. That is, the matrix of A is $[a_{ij}]_{i,j=1}^\infty$ where

$$a_{ij} = \begin{cases} \lambda_i & \text{if } i = j, \\ 0 & \text{otherwise.} \end{cases}$$

Thus, the action of A is given by

$$A\left(\sum_{n=1}^\infty x_n e_n\right) = \sum_{n=1}^\infty \lambda_n x_n e_n.$$

Then A is compact if and only if $\lambda_n \to 0$.

To see the backward implication, suppose that $\lambda_n \to 0$. We shall show that A is the limit in the operator norm of a sequence of finite rank operators. For $k \in \mathbb{N}$ define $A_k \in \mathscr{L}(H)$ by

$$A_k\left(\sum_1^\infty x_n e_n\right) = \sum_1^k \lambda_n x_n e_n.$$

A_k has rank at most k, and for an arbitrary element

$$x = \sum_1^\infty x_n e_n$$

of H we have

$$\|(A - A_k)x\|^2 = \left\|\sum_1^\infty \lambda_n x_n e_n - \sum_1^k \lambda_n x_n e_n\right\|^2$$
$$= \left\|\sum_{k+1}^\infty \lambda_n x_n e_n\right\|^2$$
$$= \sum_{k+1}^\infty |\lambda_n|^2 |x_n|^2$$
$$\leqslant \sup_{n>k} |\lambda_n|^2 \sum_{k+1}^\infty |x_n|^2$$
$$\leqslant \left(\sup_{n>k} |\lambda_n|^2\right) \|x\|^2.$$

Hence, by definition of the operator norm,

$$\|A - A_k\| \leqslant \sup_{n > k} |\lambda_n|.$$

Since the right hand side tends to zero as $k \to \infty$, we deduce that $A_k \to A$ in the operator norm. By (i) above each A_k is a compact operator, and so, by the theorem which follows, A is compact too. $\square$

8.3   Theorem   Let E, F be Banach spaces. The set of compact operators in $\mathcal{L}(E, F)$ is closed in $\mathcal{L}(E, F)$ with respect to the operator norm.

Proof. We must show that if (K_j) is a sequence of compact operators converging to a limit K in $\mathcal{L}(E, F)$ with respect to the operator norm then K is compact. Consider any bounded sequence (x_n) in E – say $\|x_n\| \leqslant M$, where $M > 0$, for all $n \in \mathbb{N}$. We must somehow use the compactness of the K_js to manufacture a convergent subsequence of (Kx_n). It can be done by a combination of two standard types of reasoning: a 'diagonal argument' and an '$\varepsilon/3$ argument'.

Since K_1 is compact there is a subsequence (x_{1n}) of (x_n) such that $(K_1 x_{1n})$ converges. Then, since K_2 is compact, there is a subsequence (x_{2n}) of (x_{1n}) such that $(K_2 x_{2n})$ converges in F. The following diagram may suggest a helpful mental picture.

$$
\begin{array}{ccccccccc}
x_1 & x_2 & x_3 & x_4 & x_5 & x_6 & x_7 & x_8 & x_9 & \cdots \\
\| & & \| & & \| & & & \| & \| & \\
x_{11} & x_{12} & & & x_{13} & & & x_{14} & x_{15} & \cdots \\
\| & & & & \| & & & & \| & \\
x_{21} & & & & x_{22} & & & & x_{23} & \cdots
\end{array}
$$

Continuing in this way we successively obtain subsequences $(x_{jn})_{n=1}^{\infty}$, $j = 1, 2, 3, \ldots$ of (x_n) with the properties that $(K_j x_{jn})_{n=1}^{\infty}$ is a convergent sequence in F for each $j \in \mathbb{N}$ and that $(x_{jn})_{n=1}^{\infty}$ is a subsequence of $(x_{in})_{n=1}^{\infty}$ for $i = 1, 2, \ldots, j-1$. We shall show that the 'diagonal subsequence' $(x_{nn})_{n=1}^{\infty}$ is such that (Kx_{nn}) converges.

Let $\varepsilon > 0$. Since $K_j \to K$ in the operator norm we can choose $p \in \mathbb{N}$ such that

$$\|K - K_p\| < \frac{\varepsilon}{3M}.$$

Now observe that $(K_p x_{nn})_{n=p}^{\infty}$ is a Cauchy sequence, for it is a subsequence of the convergent sequence $(K_p x_{pn})_{n=1}^{\infty}$. Hence there exists $n_0 > p$ such that $m, n \geqslant n_0$ imply

$$\|K_p x_{nn} - K_p x_{mm}\| < \frac{\varepsilon}{3}.$$

We are trying to show that $Kx_{nn} - Kx_{mm}$ is small knowing that K is close to K_p and $K_p x_{nn} - K_p x_{mm}$ is small. Formally, for $m, n \geqslant n_0$ we have

$$
\begin{aligned}
\|Kx_{nn} - Kx_{mm}\| &= \|Kx_{nn} - K_p x_{nn} + K_p x_{nn} - K_p x_{mm} \\
&\qquad\qquad + K_p x_{mm} - Kx_{mm}\| \\
&\leqslant \|(K - K_p)x_{nn}\| + \|K_p x_{nn} - K_p x_{mm}\| \\
&\qquad + \|(K_p - K)x_{mm}\| \\
&\leqslant \|K - K_p\|\|x_{nn}\| + \frac{\varepsilon}{3} + \|K - K_p\|\|x_{mm}\| \\
&\leqslant \frac{\varepsilon}{3M}\cdot M + \frac{\varepsilon}{3} + \frac{\varepsilon}{3M}\cdot M \\
&= \varepsilon.
\end{aligned}
$$

Thus the sequence $(Kx_{nn})_{n=1}^{\infty}$ is Cauchy in F and so convergent, by the completeness of F. Since (x_{nn}) is a subsequence of the original bounded sequence (x_n), we have proved that K is a compact operator. $\qquad\square$

8.4 Exercise Let A, B be compact operators from E to F, where E, F are Banach spaces. Prove that $A + B$ is a compact operator. $\qquad\square$

8.1 Hilbert–Schmidt operators

Example 8.2(iii) gives us a genuinely infinite-dimensional compact operator, but it is one specially made up for the purpose. Operators we encounter in the study of nature are unlikely to come in so obligingly tractable a form, and so to apply the theory we need some way of recognizing whether a particular operator is compact. If we use the definition directly then in principle we must examine all subsequences of all bounded sequences in the domain space – a tall order. There is a property stronger than compactness which is much easier to check.

8.5 Definition Let E, F be Hilbert spaces. A bounded linear operator $A: E \to F$ is said to be *Hilbert–Schmidt* if there exists a complete orthonormal sequence $(e_n)_1^{\infty}$ in E such that

$$
\sum_1^{\infty} \|Ae_n\|^2 < \infty.
$$
$\qquad\square$

8.6 Exercise The Volterra operator V on $L^2(0, 1)$ is defined by

$$
(Vf)(t) = \int_0^t f(\tau)\,d\tau, \qquad 0 < t < 1.
$$

Prove that V is a Hilbert–Schmidt operator. $\qquad\square$

8.7 *Theorem* Hilbert–Schmidt operators are compact.

Proof. Let A and (e_n) be as in Definition 8.5. We shall show that A is compact by expressing it as a norm limit of finite rank operators. Define $A_k : E \to F$ (where $k \in \mathbb{N}$) by

$$A_k\left(\sum_1^\infty x_n e_n\right) = \sum_1^k x_n A e_n$$

where

$$x = \sum_1^\infty x_n e_n \tag{8.1}$$

is an arbitrary element of E. Thus A_k agrees with A in the span of $e_1, \dots, e_k$ and is zero on the span of the remaining e_ns. The rank of A_k is at most k, and so A_k is compact. For x as in (8.1),

$$(A - A_k)x = \sum_1^\infty x_n A e_n - \sum_1^k x_n A e_n$$

$$= \sum_{k+1}^\infty x_n A e_n.$$

Hence

$$\|(A - A_k)x\| \leqslant \sum_{k+1}^\infty |x_n| \|A e_n\|$$

$$\leqslant \left\{\sum_{k+1}^\infty |x_n|^2\right\}^{1/2} \left\{\sum_{k+1}^\infty \|A e_n\|^2\right\}^{1/2}$$

$$\leqslant \|x\| \left\{\sum_{k+1}^\infty \|A e_n\|^2\right\}^{1/2}.$$

It follows from the definition of the operator norm that

$$\|A - A_k\| \leqslant \left\{\sum_{k+1}^\infty \|A e_n\|^2\right\}^{1/2}.$$

The sum inside the chain brackets is the tail of a convergent series, and hence tends to zero as $k \to \infty$. Thus $A_k \to A$ in the operator norm, and so A is compact.   □

With the aid of the Hilbert–Schmidt criterion we can describe a substantial and important class of compact integral operators.

8.8 *Theorem* Let

$$k : (c, d) \times (a, b) \to \mathbb{C}$$

be a Lebesgue measurable function such that

$$\int_c^d \int_a^b |k(t, s)|^2 \, ds \, dt < \infty. \tag{8.2}$$

Then the integral operator

$$K: L^2(a, b) \to L^2(c, d)$$

with kernel function k is a Hilbert–Schmidt operator and is therefore compact.

Here (a, b) and (c, d) can be finite or infinite intervals of the real line.

Proof. Pick any complete orthonormal sequence $(e_n)_1^\infty$ in $L^2(a, b)$. For $t \in (c, d)$ we have

$$(Ke_n)(t) = \int_a^b k(t, s) e_n(s)\, ds$$

$$= (k_t, \bar{e}_n) \tag{8.3}$$

in terms of the inner product on $L^2(a, b)$, where $k_t \in L^2(a, b)$ is defined by

$$k_t(s) = k(t, s), \qquad a < s < b \tag{8.4}$$

and $\bar{e}_n \in L^2(a, b)$ is given by

$$\bar{e}_n(s) = \overline{e_n(s)}, \qquad a < s < b.$$

Thus

$$\|Ke_n\|^2 = \int_c^d |Ke_n(t)|^2\, dt$$

$$= \int_c^d |(k_t, \bar{e}_n)|^2\, dt.$$

Now $(\bar{e}_n)_1^\infty$ is a complete orthonormal sequence in $L^2(a, b)$, and hence

$$\sum_{n=1}^\infty \|Ke_n\|^2 = \sum_{n=1}^\infty \int_c^d |(k_t, \bar{e}_n)|^2\, dt$$

$$= \int_c^d \sum_{n=1}^\infty |(k_t, \bar{e}_n)|^2\, dt \tag{8.5}$$

$$= \int_c^d \|k_t\|^2\, dt$$

$$= \int_c^d \int_a^b |k_t(s)|^2\, ds\, dt$$

$$= \int_c^d \int_a^b |k(t, s)|^2\, ds\, dt$$

$$< \infty.$$

On the face of it this concludes the proof that K is Hilbert–Schmidt, but we have been cavalier about one or two measure-theoretic points. The reader who is not familiar with the Lebesgue integral can either trust me or convince himself that the calculations are valid in the case that k is continuous on a compact rectangle $[c, d] \times [a, b]$. That was good enough for Hilbert and it is good enough for the application in Chapter 11. For the benefit of others I sketch the details. The first question is: how do we know that k_t, defined by (8.4), belongs to L^2? Indeed, does the assumption of the square integrability of k ensure that the integral operator K is well defined from L^2 to L^2? In fact it does: the Fubini–Tonelli theorem tells us that (8.2) implies that $k_t \in L^2(a, b)$ for almost all $t \in (c, d)$. Thus, for any $f \in L^2(a, b)$ we have

$$(Kf)(t) = (k_t, \bar{f}) \in \mathbb{C}, \qquad \text{almost all } t \in (a, b),$$

and hence

$$|Kf(t)| \leqslant \|k_t\| \|f\|, \qquad \text{almost all } t.$$

Thus

$$\int_c^d |Kf(t)|^2 \, dt \leqslant \|f\|^2 \int_c^d \|k_t\|^2 \, dt$$

$$= \|f\|^2 \int_c^d \int_a^b |k(t, s)|^2 \, ds \, dt.$$

This shows that $Kf \in L^2(c, d)$ and that K is a bounded operator.

The other point in the proof is the interchange of an integral and a summation in (8.5). This is justified by the monotone convergence theorem. □

Not all compact operators are Hilbert–Schmidt: the operator A on ℓ^2 with matrix

$$\text{diag}\left\{ 1, \frac{1}{\sqrt{2}}, \frac{1}{\sqrt{3}}, \frac{1}{\sqrt{4}}, \ldots \right\}$$

is compact (see Example 8.2(iii)), and if we take (e_n) to be the standard orthonormal basis of ℓ^2 we have

$$\sum \|Ae_n\|^2 = \sum \left\| \frac{1}{\sqrt{n}} e_n \right\|^2 = \sum \frac{1}{n} = \infty.$$

Now it is conceivable that there might be some other orthonormal basis of ℓ^2 for which $\sum \|Ae_n\|^2$ is finite. In fact this cannot happen: if $\sum \|Ae_n\|^2$ converges for any choice of complete orthonormal sequence (e_n) then it does so for every such choice, and moreover the value of the sum is always the same (see Problem 8.5). Hence the present A is compact but not Hilbert–Schmidt.

As all the foregoing examples suggest, one usually proves that a given operator is compact by showing that it is the norm-limit of a sequence of finite-rank operators. Is this the only way of obtaining compact operators? Are there any compact operators which are not the norm-limit of any such sequence? In the setting of Hilbert space it is fairly easy to show that the answer to the latter question is no. For Banach spaces it was a much studied open problem for several decades. It was resolved in 1973 when the Swedish mathematician P. Enflo constructed an example of a compact operator on a Banach space which was not the norm-limit of any sequence of finite rank operators.

8.2   The spectral theorem for compact Hermitian operators

Now consider the extension to compact operators of results about matrices. Practically all the deeper properties of matrices involve the notion of eigenvalue. Do compact operators have eigenvalues? Alas, they need not.

8.9 Exercise Show that the Volterra operator (defined in Exercise 8.6) has no eigenvalues.   □

Here the definition of eigenvalue is exactly as for finite-dimensional transformations. We say that λ is an eigenvalue of $A \in \mathscr{L}(E)$ if there is a non-zero vector $x \in E$ such that $Ax = \lambda x$. When this holds, x is said to be an eigenvector of A corresponding to the eigenvalue λ. Despite Exercise 8.9, eigenvalues do play a role for compact operators.

8.10 Theorem Let K be a compact Hermitian operator on a Hilbert space H. Either $\|K\|$ or $-\|K\|$ is an eigenvalue of K.

Proof. We can suppose $K \neq 0$. By Theorem 7.18

$$\|K\| = \sup_{\|x\|=1} |(Kx, x)|.$$

Hence there exists a sequence (x_n) in H of unit vectors such that

$$|(Kx_n, x_n)| \to \|K\|.$$

Since each (Kx_n, x_n) is real we can assume (by replacing (x_n) by a subsequence if necessary) that

$$(Kx_n, x_n) \to \lambda$$

where λ is either $\|K\|$ or $-\|K\|$. Since λ is real,

$$\|Kx_n - \lambda x_n\|^2 = \|Kx_n\|^2 - 2\lambda(Kx_n, x_n) + \lambda^2\|x_n\|^2$$
$$\leqslant 2\lambda^2 - 2\lambda(Kx_n, x_n).$$

The left hand side is non-negative and the right hand side tends to zero as $n \to \infty$. Hence

$$Kx_n - \lambda x_n \to 0.$$

Since K is compact there is a subsequence $(x_{n'})$ of (x_n) such that $(Kx_{n'})$ is a convergent sequence. Denote its limit by y. We have

$$Kx_{n'} \to y, \qquad Kx_{n'} - \lambda x_{n'} \to 0. \tag{8.6}$$

Subtracting, we obtain

$$\lambda x_{n'} \to y.$$

On applying the continuous operator K we find

$$\lambda K x_{n'} \to K y,$$

while, by choice of y,

$$\lambda K x_{n'} \to \lambda y.$$

Hence $Ky = \lambda y$. To conclude that λ is an eigenvalue we need only show that $y \neq 0$. We have

$$\| y \| = \lim \| \lambda x_{n'} \|$$

$$= |\lambda|$$

$$= \| K \| \neq 0.$$

Thus $y \neq 0$ and so λ is an eigenvalue of K.   □

This result gives us a good chance of extending to compact Hermitian operators those results about Hermitian matrices which involve eigenvalues. Recall that there is a particularly powerful diagonalization result for such matrices: every Hermitian matrix is unitarily equivalent to a real diagonal matrix. Another way of looking at this fact is as follows. The eigenvalues of a Hermitian matrix A are real numbers and there is an orthonormal basis of the domain of A consisting of eigenvectors of A. On taking the matrix with respect to this basis of the linear transformation defined by A we obtain a real diagonal matrix, with the eigenvalues of A as its diagonal entries. Some of these properties extend with no difficulty to infinite dimensions.

8.11 Theorem   Let A be a Hermitian operator on a Hilbert space H. All eigenvalues of A are real, and eigenvectors of A corresponding to distinct eigenvalues are orthogonal.

Proof. Suppose that λ is an eigenvalue of A and φ a corresponding eigenvector: then $A\varphi = \lambda\varphi$ and $\varphi \neq 0$. By self-adjointness of A,

$$0 = (A\varphi, \varphi) - (\varphi, A\varphi)$$

$$= (\lambda\varphi, \varphi) - (\varphi, \lambda\varphi)$$

$$= (\lambda - \bar{\lambda})\|\varphi\|^2.$$

Since $\varphi \neq 0$, $\lambda = \bar{\lambda}$, and so λ is real. Now let λ, μ be distinct eigenvalues of A

and let φ, ψ be corresponding eigenvectors, so that $A\varphi = \lambda\varphi$, $A\psi = \mu\psi$. Then

$$0 = (A\varphi, \psi) - (\varphi, A\psi)$$
$$= (\lambda\varphi, \psi) - (\varphi, \mu\psi)$$
$$= (\lambda - \bar{\mu})(\varphi, \psi).$$

Since μ is real and $\lambda \neq \mu$,

$$\lambda - \bar{\mu} = \lambda - \mu \neq 0.$$

Hence $(\varphi, \psi) = 0$, as required.   □

Observe that the statement in the theorem may hold vacuously: A need not have any eigenvalues (recall Example 7.21). If A is in addition compact then we have seen that it has at least one eigenvalue, and we can describe all possibilities for the set of eigenvalues of A.

8.12 Theorem   Let K be a compact Hermitian operator on a Hilbert space H. The set of eigenvalues of K is a set of real numbers which either is finite or consists of a countable sequence tending to zero.

Proof. Suppose that K has infinitely many eigenvalues but that they do not consist of a sequence tending to zero. Then there exists $\varepsilon > 0$ such that infinitely many eigenvalues of K have moduli greater than ε. Pick a sequence $(\lambda_n)_1^\infty$ of distinct eigenvalues of K such that $|\lambda_n| > \varepsilon, n \in \mathbb{N}$. Let φ_n be a unit eigenvector corresponding to λ_n. By 8.11, (φ_n) is an orthonormal sequence, and Pythagoras's theorem gives, for $n \neq m$,

$$\|K\varphi_n - K\varphi_m\|^2 = \|\lambda_n\varphi_n - \lambda_m\varphi_m\|^2$$
$$= |\lambda_n|^2 + |\lambda_m|^2$$
$$> 2\varepsilon^2.$$

It follows that $(K\varphi_n)$ has no Cauchy subsequence, and hence no convergent subsequence. As (φ_n) is a bounded sequence, this contradicts the compactness of K. Thus, if the eigenvalues of K are infinite in number, they must comprise a sequence tending to zero.   □

8.13 Exercise   Let $(\lambda_n)_1^\infty$ be a sequence of real numbers converging to zero. Show that there exists a compact Hermitian operator on a suitable Hilbert space whose eigenvalues consist precisely of the numbers $\{\lambda_n : n \in \mathbb{N}\}$ (recall Example 8.2(iii)).   □

8.14 Lemma   Let M be a closed linear subspace of a Hilbert space H and let M be invariant under a bounded linear operator T on H (that is, $TM \subseteq M$). Then $M^\perp$ is invariant under T^*.

Proof. Consider any $x \in M^\perp$ and $m \in M$. Then $Tm \in M$ and hence $(Tm, x) = 0$. Thus $(T^*x, m) = 0$ for all $m \in M$, so that $T^*x \in M^\perp$ for all $x \in M^\perp$. $\square$

We can now prove the promised diagonalization result for compact Hermitian operators.

8.15 Spectral theorem Let K be a compact Hermitian operator on a Hilbert space H. There exists a finite or infinite orthonormal sequence (φ_n) of eigenvectors of K, with corresponding real eigenvalues (λ_n), such that, for all $x \in H$,

$$Kx = \sum_n \lambda_n(x, \varphi_n)\varphi_n.$$

The sequence (λ_n), if infinite, tends to 0.

Proof. The reality of the (λ_n) and the final sentence follow from Theorems 8.11, 8.12.

We shall obtain the eigenvectors (φ_n) inductively. By Theorem 8.10 K has an eigenvalue $\lambda_1 = \pm \|K\|$, and we may pick a corresponding eigenvector φ_1 of unit norm. Since $\mathrm{lin}\{\varphi_1\}$ is invariant under K, its orthogonal complement H_2 is invariant under K^* by Lemma 8.14, and $K^* = K$. Thus $KH_2 \subseteq H_2$ and so we may define K_2 to be the restriction of K to H_2. Clearly K_2 is compact, and if $x, y \in H_2$ then

$$(K_2^*x, y) = (x, K_2 y) = (x, Ky)$$
$$= (K^*x, y) = (Kx, y).$$

Since $Kx \in H_2$, it follows that $Kx = K_2^*x$, all $x \in H_2$. Thus K_2^* is the restriction of K to H_2: that is, $K_2^* = K_2$. Thus K_2 is a compact Hermitian operator on H_2, and so we may apply Theorem 8.10 again to obtain an eigenvalue $\lambda_2 = \pm \|K_2\|$ of K_2, and hence of K. Choose a unit eigenvector φ_2 of K_2 corresponding to λ_2.

Inductively, suppose we have constructed mutually orthogonal unit eigenvectors $\varphi_1, \ldots, \varphi_n$ of K with corresponding eigenvalues $\lambda_1, \ldots, \lambda_n$ such that $|\lambda_j| = \|K_j\|$, $1 \leqslant j \leqslant n$, where $K_1 = K$ and K_j is the restriction of K to $\{\varphi_1, \ldots, \varphi_{j-1}\}^\perp$, $2 \leqslant j \leqslant n$. Then let $H_{n+1} = \{\varphi_1, \ldots, \varphi_n\}^\perp$ and let K_{n+1} be the restriction of K to H_{n+1} (this is possible since, by Lemma 8.14, H_{n+1} is invariant under K). K_{n+1} is a compact Hermitian operator on H_{n+1}, and so by Theorem 8.10 K_{n+1} has an eigenvalue $\lambda_{n+1} = \pm \|K_{n+1}\|$, with corresponding unit eigenvector $\varphi_{n+1} \in H_{n+1}$. Then $\varphi_1, \varphi_2, \ldots, \varphi_{n+1}$ are mutually orthogonal unit eigenvectors, and their corresponding eigenvalues $\lambda_1, \ldots, \lambda_{n+1}$ satisfy $|\lambda_j| = \|K_j\|$, $1 \leqslant j \leqslant n + 1$.

We continue this inductive construction as long as $K_n \neq 0$. If we encounter an m such that $K_m = 0$ the construction stops and we have, for an arbitrary $x \in H$,

$$x - \sum_{j=1}^{m-1} (x, \varphi_j)\varphi_j \in H_m.$$

Denoting the left hand side by y, we have

$$0 = K_m y = Ky$$
$$= Kx - \sum_{j=1}^{m-1} (x, \varphi_j)K\varphi_j.$$

That is, for any $x \in H$,

$$Kx = \sum_{j=1}^{m-1} \lambda_j(x, \varphi_j)\varphi_j,$$

and the assertion of Theorem 8.15 is valid with a finite sum.

Alternatively, $K_n \neq 0$ for all $n \in \mathbb{N}$. Consider an arbitrary $x \in H$ and let

$$y_n = x - \sum_{j=1}^{n-1} (x, \varphi_j)\varphi_j.$$

Then $y_n \in H_n$. Applying Pythagoras' theorem to the relation

$$x = y_n + \sum_{j=1}^{n-1} (x, \varphi_j)\varphi_j$$

we see that

$$\|x\|^2 = \|y_n\|^2 + \sum_{j=1}^{n-1} |(x, \varphi_j)|^2,$$

so that

$$\|y_n\| \leq \|x\|.$$

Now

$$\|Ky_n\| = \|K_n y_n\| \leq \|K_n\| \|y_n\|$$
$$\leq |\lambda_n| \|x\|.$$

That is,

$$\left\| Kx - \sum_{j=1}^{n-1} \lambda_j(x, \varphi_j)\varphi_j \right\| \leq |\lambda_n| \|x\|.$$

By Theorem 8.12, $\lambda_n \to 0$ as $n \to \infty$. Hence

$$\sum_{j=1}^{\infty} \lambda_j(x, \varphi_j)\varphi_j = Kx$$

for all $x \in H$. □

The orthonormal sequence (φ_n) promised by this theorem need not be complete: indeed, if H is not separable then no orthonormal sequence is complete in H. It is often helpful to think in terms of orthonormal bases, so we note that the (φ_n) above can be extended to a basis in the separable case.

8.16 Corollary Let K be a compact Hermitian operator on a separable infinite-dimensional Hilbert space H. There exists a complete orthonormal sequence $(e_n)_{n \in \mathbb{N}}$ consisting of eigenvectors of K. For any $x \in H$,

$$Kx = \sum_{n=1}^{\infty} \lambda_n (x, e_n) e_n,$$

where λ_n is the eigenvalue of K corresponding to e_n.

Proof. By the spectral theorem there is a finite or infinite orthonormal sequence (φ_n) such that, for all $x \in H$,

$$Kx = \sum_n \lambda_n (x, \varphi_n) \varphi_n. \tag{8.7}$$

Any term in which $\lambda_n = 0$ can be omitted from this sum, so we may suppose each $\lambda_n \neq 0$. Let (ψ_m) be a complete orthonormal sequence in the Hilbert space $\operatorname{Ker} K$, the kernel of K. Then each ψ_m is an eigenvector of K corresponding to the eigenvalue 0. As φ_n is an eigenvector corresponding to a different eigenvalue, φ_n is orthogonal to ψ_m by Theorem 8.11. Thus $\{\varphi_n\} \cup \{\psi_m\}$ is a countable orthonormal set in H. For any $x \in H$, (8.7) implies that

$$x - \sum_n (x, \varphi_n) \varphi_n \in \operatorname{Ker} K,$$

and since (ψ_m) is an orthonormal basis of $\operatorname{Ker} K$,

$$x - \sum_n (x, \varphi_n) \varphi_n = \sum_m (x, \psi_m) \psi_m.$$

Thus $\{\varphi_n\} \cup \{\psi_m\}$ is a complete orthonormal set in H. Since $\dim H = \infty$ this set is infinite, so it may be indexed by $\mathbb{N}$, as $(e_n)_{n \in \mathbb{N}}$. $\qquad \square$

The corollary means that we can choose an orthonormal basis of H so that the corresponding matrix of K is diagonal, with real diagonal entries tending to zero. This often makes it easy to visualize properties of a general compact Hermitian operator.

8.17 Exercise Let K be a compact Hermitian operator on a Hilbert space H and suppose that $(Kx, x) \geqslant 0$ for all $x \in H$. Show that all eigenvalues of K are non-negative, and prove that there exists a Hermitian operator A on H such that $A^2 = K$. $\qquad \square$

The spectral theorem for compact Hermitian operators was essentially proved by Hilbert in a paper published in 1906, though his viewpoint and terminology were different. E. Schmidt, E. Fischer and F. Riesz then (around 1908) introduced the geometric approach which made it all so much easier and more elegant. One could say that this was the point at which functional analysis crystallized, out of the 'confluence of geometry, topology and analysis', in the words of J. Dieudonné (1981, Ch. 5). The work of many mathematicians had led to this development. Indeed, the major causative influence which Dieudonné identifies is the tendency to increasing abstraction, which is rather a property of mathematical culture as a whole. Another important factor was the timely appearance of Lebesgue's theory of integration, to replace the 'horrible and useless so-called Riemann integral' (Dieudonné, 1981, p. 119).

The spectral theorem for compact Hermitian operators was the first substantial result in the branch of functional analysis known as operator theory. It was soon followed by a satisfactory spectral theorem (due to Hilbert and F. Riesz) for Hermitian operators which are bounded but not compact. Such operators need not have eigenvalues, and so one has to use a more subtle notion involving measure theory. This line of investigation was completed by M. H. Stone and von Neumann in the late 1920s with a theory for unbounded Hermitian operators. There is also a beautiful theory of compact non-Hermitian operators: this is necessarily less simple than the case we have studied, but is still a powerful tool. This theory is due to F. Riesz in 1918 (the reason for the insistence on F., for Frigyes, is that he had a brother, Marcel, who was also a mathematician of great power).

Many operators are neither compact nor Hermitian, so operator theorists have had plenty to occupy them over the past eight decades. There is an enormous, rich and extensive theory covering many classes of operators. There are half a dozen major journals devoted primarily to operators, their ramifications in other branches of mathematics and their applications in physics and engineering. These and more general journals publish thousands of new theorems every year about operators. The reader may find it instructive to dip – briefly! – into a recent issue of the *Journal of Functional Analysis* or the *Journal of Operator Theory* or *Integral Equations and Operator Theory*. Practically every problem in analysis can be reformulated in operator-theoretic terms, so it will surely never come about that there is no more to say about operators. A recent triumph of such a re-formulation was the solution, with the aid of operator-theoretic methods, of one of the longest-standing and most intensively studied problems of complex analysis, the 'Bieberbach conjecture'. This was achieved in 1984 by L. de Branges.

8.3   Problems

8.1. Let E be a Banach space and let $A, B, C \in \mathcal{L}(E)$. Show that if B is compact then so is ABC.

8.2. Let E be an infinite-dimensional Hilbert space and let $A, B \in \mathcal{L}(E)$.

Which of the following are true?

    (a) If AB is compact then either A or B is compact.

    (b) If $A^2 = 0$ then A is compact.

    (c) If $A^n = I$ for some $n \in \mathbb{N}$ then A is not compact.

8.3. Let E, F be normed spaces and let $A \in \mathscr{L}(E, F)$. Prove that A is compact if and only if A maps the unit ball of E onto a relatively compact set in F – in other words, if the closure of the set

$$\{Ax : x \in E, \|x\| \leqslant 1\}$$

is compact in F.

8.4. Let $A: W[-\pi, \pi] \to L^2(-\pi, \pi)$ be the injection operator (so that $Af = f$). Prove that A is Hilbert–Schmidt (assume the result of Problem 5.9).

8.5. Let $(e_n)_1^\infty, (f_m)_1^\infty$ be complete orthonormal sequences in Hilbert spaces H, K respectively and let $A \in \mathscr{L}(H, K)$. Show that

$$\sum_{n=1}^{\infty} \|Ae_n\|^2 = \sum_{m=1}^{\infty} \|A^*f_m\|^2.$$

Deduce that the quantity $\sum_{n=1}^{\infty} \|Ae_n\|^2$ has the same value for every choice of the complete orthonormal sequence (e_n) in H.

    The quantity $\{\sum_{n=1}^{\infty} \|Ae_n\|^2\}^{1/2}$, if finite, is called the *Hilbert–Schmidt norm* of A. Show that it equals

$$\left\{ \sum_{i,j=1}^{\infty} |a_{ij}|^2 \right\}^{1/2}$$

if A has matrix $[a_{ij}]$ with respect to any pair of complete orthonormal sequences in H and K.

8.6. Show that, for any Hilbert–Schmidt operator A, the operator norm of A is less than or equal to the Hilbert–Schmidt norm of A.

8.7. Can a multiplication operator on $L^2(a, b)$ be Hilbert–Schmidt? Compact?

8.8. Is the operator

$$\frac{\mathrm{d}}{\mathrm{d}x} : W[-\pi, \pi] \to L^2(-\pi, \pi)$$

compact?

8.9. Show that the square integrability condition (8.2) is also necessary for the operator K in Theorem 8.8 to be Hilbert–Schmidt.

8.10. The integral operator K on $L^2(0, \infty)$ is defined by

$$(Kf)(x) = \int_0^\infty k(x + \tau) f(\tau) \, \mathrm{d}\tau,$$

where $k:(0, \infty) \to \mathbb{C}$ is a continuous function such that

$$M = \int_0^\infty \tau |k(\tau)|^2 \, d\tau < \infty.$$

Prove that K is Hilbert–Schmidt, with Hilbert–Schmidt norm $M^{1/2}$.

8.11. Let K be a compact Hermitian operator on a Hilbert space H and let the kernel of K be $\{0\}$. Show that there is a sequence (K_n) of bounded linear operators on H such that

$$K_n K x \to x, \qquad K K_n x \to x \qquad \text{as } n \to \infty.$$

Can the K_n be chosen so that $K_n K \to I$ in the operator norm?

8.12. Show that if some positive power of a compact Hermitian operator K is zero then K is itself zero.

8.13. Let K be a compact Hermitian operator on a Hilbert space H. Show that if K has infinite rank then the range of K is not closed in H.

8.14. Show that a compact Hermitian operator is the limit with respect to the operator norm of a sequence of finite rank operators.

8.15. Let K be a compact operator on a Hilbert space. Prove that the closure of the range of K is a separable Hilbert space (recall Problem 7.29).

8.16. Let H be a separable Hilbert space and let K be a compact operator on H. Let $(\varphi_j)_1^\infty$ be a complete orthonormal sequence of eigenvectors of K^*K and let $(\lambda_j)_1^\infty$ be the sequence of corresponding eigenvalues. Show that the formula

$$U\left(\sum_{j=1}^\infty x_j \varphi_j\right) = \sum_{j=1}^\infty x_j \mu_j K \varphi_j,$$

where

$$\mu_j = \begin{cases} \lambda_j^{-1/2} & \text{if } \lambda_j > 0 \\ 0 & \text{if } \lambda_j = 0, \end{cases}$$

defines U as a bounded linear operator on H. What is $\|U\|$? Show further that $K = U(K^*K)^{1/2}$. Deduce, using Problems 8.1 and 8.14, that K is the limit with respect to the operator norm of a sequence of finite rank operators.

8.17. Put together the results of Problems 8.15 and 8.16 to show that a compact operator on an arbitrary Hilbert space is the limit with respect to the operator norm of a sequence of finite rank operators.

9

Sturm–Liouville systems

There is no shortage of essential uses of Hilbert space theory in mathematics and in pure and applied science, but setting up the background for any of them is lengthy. We shall content ourselves with two applications which illustrate Hilbert spaces in action in mainstream mathematics and also in science and engineering. In the next three chapters we shall develop the mathematical theory which justifies the method of 'separation of variables'. This is one of the commonest approaches to the solution of linear partial differential equations, and is in routine use by scientists and working engineers. The topic thus has practical importance, and it also has historical significance since it was from the study of the differential and integral equations associated with such problems that functional analysis emerged.

Sturm–Liouville systems are second-order linear differential equations with boundary conditions of a particular type, and they usually arise from separation of variables in partial differential equations which represent physical systems. As an illustration we analyse small planar oscillations of a hanging chain.

9.1   Small oscillations of a hanging chain

A uniform heavy flexible chain of length L is freely suspended at one end and hangs under gravity. The chain is displaced slightly, in a vertical plane, and released from rest. The problem is to describe the subsequent motion of the chain. We are really concerned here with the mathematical analysis which is required, but for completeness let us give a brief heuristic derivation of the governing equations.

Clearly all the motion will occur in the vertical plane of the initial displacement of the chain. Take as origin O the point occupied by the lower end of the chain when it hangs at rest (Diagram 9.1) and take the positive Ox axis to be vertically upwards. Let the horizontal displacement of the chain at time t and height x above O be $u(x, t)$ (choose one of the two relevant horizontal directions as the *positive u* axis). The tension T in the chain at height x is due to the weight of the chain beneath that point. Since we are concerned only with small oscillations, the length of this portion of the chain is approximately x, and so we can suppose that T is proportional to x. For simplicity we shall assume that the density of the chain and the units of measurement are such that $T = x$. To obtain an equation of motion consider the short segment of the chain between heights x and $x + \delta x$ (Diagram 9.2). The horizontal component of tension in this element is roughly $x \, \partial u / \partial x$, and so the change in this component over the length of the segment is approximately

$$\delta x \cdot \frac{\partial}{\partial x} \left(x \frac{\partial u}{\partial x} \right).$$

Diagram 9.1

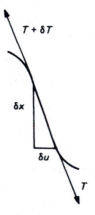

Diagram 9.2

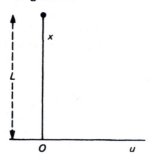

The mass of this segment is $\rho\,\delta x$, where ρ is the mass/unit length of the chain. For simplicity let us suppose that $\rho = 1$. Then Newton's second law tells us that

$$\delta x \cdot \frac{\partial}{\partial x}\left(x\,\frac{\partial u}{\partial x}\right) = \delta x \cdot \frac{\partial^2 u}{\partial t^2}\,.$$

We therefore take the equation of motion of the chain to be

$$\frac{\partial^2 u}{\partial t^2} = \frac{\partial}{\partial x}\left(x\,\frac{\partial u}{\partial x}\right). \tag{9.1}$$

If the initial displacement of the chain at height x is $u_0(x)$ and if we assume that the chain is released from rest at time $t = 0$ then we obtain the following boundary and initial conditions:

$$u(L, t) = 0, \qquad 0 \leqslant t < \infty, \tag{9.2}$$

$$u(x, 0) = u_0(x), \qquad 0 \leqslant x \leqslant L, \tag{9.3}$$

$$\frac{\partial u}{\partial t}(x, 0) = 0, \qquad 0 \leqslant x \leqslant L, \tag{9.4}$$

$$\sup|u(x, t)| < \infty, \qquad 0 \leqslant x \leqslant L,\, 0 \leqslant t < \infty. \tag{9.5}$$

The traditional way to try to solve this system of equations for the unknown function $u(x, t)$ is to use the separation of variables. That is, we try to express the solution as a sum of *normal modes*. A normal mode of the system is a function of the form

$$u(x, t) = f(x)g(t) \tag{9.6}$$

which satisfies the differential equation (9.1) and the boundary and boundedness conditions (9.2), (9.4), (9.5), though not necessarily the initial condition (9.3). On substituting (9.6) in the differential equation (9.1) and dividing through by $f(x)g(t)$ we find that f and g must satisfy

$$\frac{g''(t)}{g(t)} = \frac{f'(x)}{f(x)} + x\,\frac{f''(x)}{f(x)}$$

whenever $f(x)g(t) \neq 0$. Since the left hand side is independent of x and the right hand side is independent of t, both sides must be constant. Denote this constant value by $-\lambda$. We have shown that if (9.6) is a normal mode of the system then there exists $\lambda \in \mathbb{R}$ such that

$$\frac{d^2 g}{dt^2}(t) + \lambda g(t) = 0, \qquad 0 \leqslant t < \infty, \tag{9.7}$$

$$\frac{d}{dx}\left(x\,\frac{df}{dx}\right) + \lambda f(x) = 0, \qquad 0 \leqslant x \leqslant L, \tag{9.8}$$

and on adding the boundary conditions we obtain the further relations

$$f(L) = 0, \tag{9.9}$$

$$\sup_{t \geqslant 0} |g(t)| < \infty \tag{9.10}$$

which must hold if $f(x)g(t)$ is not the trivial solution of (9.1), identically equal to zero. Conversely, if f and g are twice differentiable functions satisfying (9.7)–(9.10) then $u = f(x)g(t)$ is a normal mode of the system. We can therefore find all normal modes by solving (9.7)–(9.10). Now (9.7) is the equation of simple harmonic motion, so that we can write down its general solution explicitly and immediately perceive that it has a bounded solution only when $\lambda \geqslant 0$. And if we change the independent variable from x to

$$\xi = 2\sqrt{(\lambda x)}$$

then (9.8) becomes (if $f_1(\xi) = f(x)$)

$$\frac{d^2 f_1}{d\xi^2} + \frac{1}{\xi}\frac{df_1}{d\xi} + f_1 = 0,$$

which is Bessel's equation of order 0 (see Sneddon, 1961). The general bounded solution of this ordinary differential equation is

$$f_1(\xi) = cJ_0(\xi)$$

where c is an arbitrary constant and J_0 is the Bessel function of order 0. That is,

$$f(x) = cJ_0(2\sqrt{(\lambda x)}).$$

Thus the boundary condition (9.9) is satisfied if and only if $\lambda \geqslant 0$ is such that

$$J_0(2\sqrt{(\lambda L)}) = 0.$$

The zeros of J_0 in $\mathbb{R}^+$ form an infinite sequence tending to ∞, so that they can be enumerated as

$$0 < z_1 < z_2 < \cdots$$

Then the non-negative values of λ for which (9.8) and (9.9) have a solution are given by

$$\lambda_j = \frac{z_j^2}{4L}, \qquad j = 1, 2, \ldots,$$

and the general solution corresponding to λ_j is

$$f(x) = c_j J_0(2\sqrt{(\lambda_j x)})$$

$$= c_j J_0\left(z_j \sqrt{\left(\frac{x}{L}\right)}\right).$$

The solution of (9.7), (9.10) corresponding to $\lambda = \lambda_j$ is of course

$$g(t) = a_j \cos \sqrt{(\lambda_j)} t + b_j \sin \sqrt{(\lambda_j)} t$$

$$= a_j \cos \frac{z_j t}{2\sqrt{L}} + b_j \sin \frac{z_j t}{2\sqrt{L}} .$$

Hence the general normal mode of the system (9.1)–(9.5) is

$$u_j(x, t) = J_0\left(z_j \sqrt{\left(\frac{x}{L}\right)}\right)\left(a_j \cos \frac{z_j t}{2\sqrt{L}} + b_j \sin \frac{z_j t}{2\sqrt{L}}\right) \tag{9.11}$$

where z_j is the jth zero of the Bessel function J_0 and a_j, b_j are arbitrary constants. If the solution of the original system (9.1)–(9.5) is expressible as a sum of normal modes then we must have

$$u(x, t) = \sum_{j=1}^{\infty} u_j(x, t) \tag{9.12}$$

where a_j, b_j in (9.11) are chosen to make the initial conditions (9.3), (9.4) hold. That is, we should have

$$u_0(x) = \sum_{j=1}^{\infty} a_j J_0\left(z_j \sqrt{\left(\frac{x}{L}\right)}\right), \qquad 0 \leqslant x \leqslant L,$$

$$0 = \sum_{j=1}^{\infty} \frac{b_j z_j}{2\sqrt{L}} J_0\left(z_j \sqrt{\left(\frac{x}{L}\right)}\right), \qquad 0 \leqslant x \leqslant L.$$

We can satisfy these equations by taking all b_j equal to 0 and the a_js to be the Fourier coefficients of u_0 with respect to an appropriate orthonormal sequence in $L^2(0, L)$, and so finally we obtain an infinite series expansion for the unknown function $u(x, t)$ in (9.1)–(9.5).

This method, the separation of variables, is one of the commonest ways by which applied mathematicians solve the standard equations of mathematical physics. It can be found in all text books on advanced mathematical methods for scientists and engineers, and the reader will doubtless have met it before. The purpose of the present discussion is to scrutinize the technique more critically. For the working engineer faced with a heat transfer problem the verdict of experience is sufficient: separation of variables works. The mathematician, however, will notice several questionable steps in the account above. What guarantee is there that the general solution *is* expressible as an infinite sum of normal modes? In what sense does the series (9.12) converge? Are there any restrictions on the initial displacement function u_0 which can be allowed? In the example under discussion we do at least know that there is an infinite sequence of linearly independent normal modes, since we can write them down in terms of a special function with known properties. For slightly more complicated problems this will often fail to be true – for instance, if the chain is not of uniform density. In such an example, can we be sure that there are any normal modes at all, never mind enough to express the general solution? These questions were answered by C. Sturm and J. Liouville in the 1830s, long before the advent of Hilbert space theory. Their results were precursors of operator theory, but from our present viewpoint can be most naturally obtained as consequences of the spectral theorem for compact Hermitian operators. As often happens, ingenious special calculations led eventually to a powerful general theory which allowed mathematicians to go far beyond the original results.

9.1 Definition   A *regular Sturm–Liouville system* consists of a differential equation on a finite interval $[a, b]$ of $\mathbb{R}$, together with boundary conditions, of the following type

$$\text{RSL:} \quad \frac{d}{dx}\left(p\frac{df}{dx}\right) + (\lambda\rho + q)f = 0, \tag{9.13}$$

$$\left.\begin{array}{l} \alpha f(a) + \alpha' f'(a) = 0, \\ \beta f(b) + \beta' f'(b) = 0 \end{array}\right\} \tag{9.14}$$

where

    p, ρ and q are continuous real-valued functions on $[a, b]$;
    p and ρ are positive on $[a, b]$;
    p' exists and is continuous on $[a, b]$;
    $\alpha, \alpha', \beta, \beta'$ are real constants and the trivial boundary conditions
    $\alpha = \alpha' = 0$, $\beta = \beta' = 0$ are both excluded.   □

In our example above the equation for f is of Sturm–Liouville type, with $p(x) = x$, $q = 0$ and $\rho = 1$ on the interval $[0, L]$. However, the system consisting of equations (9.8) and (9.9) is not a regular Sturm–Liouville system since p vanishes at an end point of $[0, L]$ and there is no boundary condition at that end point.

9.2 Definition   Let p, ρ and q be functions on a finite interval $[a, b]$ satisfying all the hypotheses in Definition 9.1 except that p is only supposed positive on the *open* interval (a, b), vanishing at one or both end points. The *singular Sturm–Liouville system* corresponding to these functions consists of the differential equation (9.13) on the (half open or open) interval on which p is strictly positive, together with suitable boundary conditions. These conditions are of the form

    (i) f is bounded on (a, b);
    (ii) at an end point at which p does not vanish, if there is one, p satisfies
        a condition of the type (9.14).   □

For concreteness let us write down an example of a singular system. Suppose that $p(a) = 0$ but $p(b) > 0$. The corresponding system is then

$$\text{SSL:} \quad \frac{\mathrm{d}}{\mathrm{d}x}\left(p\frac{\mathrm{d}f}{\mathrm{d}x}\right) + (\lambda\rho + q)f = 0, \quad a < x \leqslant b; \tag{9.15}$$

$$\beta f(b) + \beta' f'(b) = 0 \tag{9.16a}$$

where β, β' are real constants, not both zero;

$$f \text{ is bounded on } (a, b). \tag{9.16b}$$

It is obvious what the singular systems corresponding to the other two possibilities are $(p(a) > 0, \; p(b) = 0$ and $p(a) = p(b) = 0)$.

The differential operator on the left hand side of (9.13) may look unnecessarily special: why not consider a general linear second-order differential equation (DE)? In fact there is no loss of generality in studying (9.13). For example, the equation

$$xf'' + (1 + 3x)f' + x^2f = 0$$

can be multiplied through by e^{3x} to give

$$(x\,e^{3x}f')' + x^2\,e^{3x}f = 0.$$

See Problem 9.3.

Violins and drums furnish further examples of Sturm–Liouville systems (Problems 9.1, 9.2). To a first approximation their motions are described by the wave equation (in 1 and 2 space dimensions) with appropriate boundary conditions. On separating variables one obtains a regular and a singular Sturm–Liouville system respectively. The one-dimensional Schrödinger equation for a particle in quantum mechanics is a Sturm–Liouville-type equation, but is defined on an infinite interval.

9.2   Eigenfunctions and eigenvalues

As our example showed, a Sturm–Liouville system may have non-trivial solutions only for special values of the scalar parameter λ.

9.3 Definition An *eigenfunction* of a (regular or singular) Sturm–Liouville system corresponding to a scalar λ is a non-zero twice differentiable function f which satisfies the differential equation (9.13) and the accompanying boundary conditions. A scalar λ is said to be an *eigenvalue* of the system if there exists an eigenfunction of the system corresponding to λ. □

9.4 Example On $[0, \pi]$ consider the regular Sturm–Liouville system

$$f'' + \lambda f = 0, \tag{9.17}$$

$$f(0) = f(\pi) = 0. \tag{9.18}$$

Let us find all eigenvalues and eigenfunctions. Suppose first that $\lambda < 0$. Let $\mu = |\lambda|^{1/2}$: then the general solution of (9.17) is

$$f(x) = A\,e^{\mu x} + B\,e^{-\mu x},$$

and this function satisfies the boundary conditions (9.18) if and only if

$$A + B = 0,$$
$$A\,e^{\mu\pi} + B\,e^{-\mu\pi} = 0.$$

Since $\mu > 0$, $e^{\mu\pi} \neq \pm 1$ and hence the only solution of these equations is

$$A = B = 0.$$

This shows that there is no non-zero solution of (9.17) and (9.18) when $\lambda < 0$: in other words (9.17)–(9.18) has no negative eigenvalues. Similar reasoning shows that 0 is not an eigenvalue. For $\lambda > 0$, the general solution of (9.17) is

$$f(x) = A\cos(\sqrt{\lambda}\,x) + B\sin(\sqrt{\lambda}\,x).$$

The boundary conditions (9.18) become

$$A = 0, \qquad B\sin(\sqrt{\lambda}\,\pi) = 0.$$

For f to be an eigenfunction, B must be non-zero, and hence $\sin(\sqrt{\lambda}\,\pi) = 0$, which is satisfied if and only if $\sqrt{\lambda} \in \mathbb{N}$. Thus the eigenvalues of the system comprise the set $\{k^2 : k \in \mathbb{N}\}$, and the eigenfunctions corresponding to k^2 are the non-zero scalar multiples of the function $\sin kx$.   □

Observe that in both the examples we have seen there is an infinite sequence of eigenvalues, all real, tending to infinity.

9.5 Exercise   Find all eigenvalues and eigenfunctions of the regular Sturm–Liouville system

$$f'' + \lambda f = 0,$$
$$f'(0) = f'(\pi) = 0$$

on $[0, \pi]$.   □

The examples in 9.4 and 9.5 are the archetypal Sturm–Liouville systems. Fourier was successful in solving heat conduction problems because these systems do have eigenvalues and eigenfunctions which can be found so easily and explicitly in terms of familiar functions. Even this special case surprised many mathematicians of the time: it must have looked hopeless to try to justify the method for general p, ρ and q, when it was impossible to find the eigenvalues and eigenfunctions explicitly. In fact it was a remarkably short time before Sturm and Liouville showed that there are always enough eigenvalues and eigenfunctions and so justified the method of separation of variables for a wide range of physical problems.

Before proving the existence results let us derive some comparatively simple properties of eigenvalues and eigenfunctions. We shall denote by L the differential operator appearing in the Sturm–Liouville differential equation. Thus, for the regular system RSL we define $D(L)$ to be the space of $\mathbb{C}$-valued functions f on $[a, b]$ such that (i) f'' exists and belongs to $L^2(a, b)$, and (ii) f satisfies the boundary conditions (9.14). $L: D(L) \to L^2(a, b)$ is given by

$$Lf = \frac{d}{dx}\left(p\frac{df}{dx}\right) + qf. \tag{9.19}$$

For the singular system SSL we need only replace (9.14) by (9.16) in condition (ii).

9.6 Lagrange's identity For any $u, v \in D(L)$,

$$uLv - vLu = (p(uv' - vu'))'.$$

The proof is an easy exercise. $\qquad\qquad\square$

9.7 Self-adjointness property For any $u, v \in D(L)$,

$$(Lu, v) = (u, Lv)$$

in the inner product of $L^2(a, b)$.

Proof. If v is in $D(L)$ then so is $\bar{v}$: this is because the constants α etc. occurring in the boundary conditions (9.14) or (9.16) are real ($\bar{v}$ is of course the function whose value at x is the complex conjugate of $v(x)$). Since, further, p, ρ and q are real-valued,

$$\overline{Lv} = L\bar{v}.$$

Thus

$$(Lu, v) - (u, Lv) = \int_a^b \bar{v}Lu - uL\bar{v}\,dx$$

$$= [p(u\bar{v}' - \bar{v}u')]_a^b, \tag{9.20}$$

by 9.6 and the fundamental theorem of calculus. Now the right hand side of this equation vanishes at both end points of $[a, b]$, whether the system be regular or singular. If $p(a) = 0$ the conclusion is immediate, while if $p(a) > 0$, then u and $\bar{v}$ satisfy a boundary condition of the type (9.14) at a, by the definition of $D(L)$. That is,

$$\begin{bmatrix} u(a) & u'(a) \\ \bar{v}(a) & \bar{v}'(a) \end{bmatrix}\begin{bmatrix} \alpha \\ \alpha' \end{bmatrix} = 0$$

where $(\alpha, \alpha') \neq (0, 0)$. Hence

$$u(a)\bar{v}'(a) - \bar{v}(a)u'(a) = 0.$$

Similar reasoning applies to the other end point, and so the right hand side of (9.20) is zero. $\qquad\qquad\square$

9.8 Theorem　The eigenvalues of a Sturm–Liouville system are real numbers.

Proof. Let λ be an eigenvalue of a Sturm–Liouville system (RSL or SSL) and let f be a corresponding eigenfunction. This means $f \neq 0$ and

$$Lf = -\lambda \rho f.$$

By self-adjointness,

$$0 = (Lf, f) - (f, Lf)$$

$$= (-\lambda \rho f, f) - (f, -\lambda \rho f)$$

$$= (\bar{\lambda} - \lambda) \int_a^b \rho(x) |f(x)|^2 \, dx.$$

Since $\rho > 0$ on $[a, b]$ and $f \neq 0$ the integral is positive. Hence $\lambda = \bar{\lambda}$.　□

9.3　Orthogonality of eigenfunctions

9.9　*Orthogonality theorem*　Let u, v be eigenfunctions of one of the Sturm–Liouville systems RSL or SSL, corresponding to distinct eigenvalues. Then $\rho^{1/2} u$ is orthogonal to $\rho^{1/2} v$.

Proof. By hypothesis,

$$Lu = -\lambda \rho u, \qquad Lv = -\mu \rho v$$

for some distinct scalars λ, μ. By the preceding theorem, $\lambda, \mu \in \mathbb{R}$. By the self-adjointness property, $(Lu, v) = (u, Lv)$. Thus

$$0 = (Lu, v) - (u, Lv)$$

$$= (-\lambda \rho u, v) - (u, -\mu \rho v)$$

$$= (\mu - \lambda) \int_a^b \rho(x) u(x) \bar{v}(x) \, dx$$

$$= (\mu - \lambda)(\rho^{1/2} u, \rho^{1/2} v).$$

Since $\mu - \lambda \neq 0$, $\rho^{1/2} u \perp \rho^{1/2} v$.　□

　　By quite simple arguments we have come some way towards 'diagonalizing' the differential operator L. For simplicity, suppose for the moment that $\rho = 1$. Corresponding to each eigenvalue of L we can pick a unit eigenvector, and the collection of these eigenvectors forms an orthonormal system in $L^2(a, b)$. We should like to know that this system is in fact a complete orthonormal sequence in $L^2(a, b)$: then the matrix of L with respect to the sequence would indeed be diagonal and the analysis of the Sturm–Liouville system would become correspondingly straightforward. At the moment we do not even know whether there are *any* eigenvalues of the system. The fact that there are enough eigenfunctions is the substance of Sturm–Liouville theory. We can at least show that there are not too many eigenvalues. This will be important in Chapter 11.

9.10 Theorem   Not every real number is an eigenvalue of RSL.

Proof. We have to use the notion of *countability*. A set is said to be countable if it is finite or its elements can be put into 1–1 correspondence with the natural numbers. We need two facts about such sets: the union of a countable sequence of countable sets is countable, and $\mathbb{R}$ is not countable (see Binmore, 1981, Book 1, Chapter 12).

Let (e_n) be a complete orthonormal sequence in $L^2(a, b)$. By Theorem 5.1 there is such a sequence in $L^2(-\pi, \pi)$, and we can obtain one in $L^2(a, b)$ by suitably scaling the independent variable. Now suppose that every $\lambda \in \mathbb{R}$ is an eigenvalue of RSL. By Theorem 9.9 there is an orthonormal system $(f_\lambda)_{\lambda \in \mathbb{R}}$ in $L^2(a, b)$. For each $n \in \mathbb{N}$ let $E_n = \{\lambda \in \mathbb{R} : (e_n, f_\lambda) \neq 0\}$. We can write

$$E_n = \bigcup_{m=1}^{\infty} \left\{ \lambda \in \mathbb{R} : |(e_n, f_\lambda)| \geqslant \frac{1}{m} \right\}.$$

Bessel's inequality (Theorem 4.9) shows that

$$\{\lambda \in \mathbb{R} : |(g, f_\lambda)| \geqslant c\}$$

is finite for every $g \in L^2(a, b)$ and every $c > 0$. Hence E_n is countable for each $n \in \mathbb{N}$. Thus $\bigcup_n E_n$ is a countable subset of $\mathbb{R}$, and so a proper subset. If $\lambda \in \mathbb{R} \backslash \bigcup_n E_n$ then $f_\lambda \perp e_n$ for all $n \in \mathbb{N}$, but $f_\lambda \neq 0$, contradicting completeness of (e_n). $\square$

9.4   Problems

9.1.   The small vibrations of a taut elastic string can be roughly described by the equations

$$\frac{\partial^2 u}{\partial t^2} = a^2 \frac{\partial^2 u}{\partial x^2}, \qquad 0 \leqslant x \leqslant L, \, t \geqslant 0,$$

$$u(0, t) = u(L, t) = 0, \qquad t \geqslant 0.$$

Here L is the length of the string, $u(x, t)$ denotes the transverse displacement at time t of a point distant x from one end of the string and a is a positive constant. Separate variables to obtain a regular Sturm–Liouville system on $[0, L]$. Find the eigenvalues and eigenfunctions of this system.

9.2. The small, centrally symmetric vibrations of a stretched uniform circular membrane, fixed round its perimeter, are approximately described by the equations

$$\frac{\partial^2 u}{\partial t^2} = a^2 \left(\frac{\partial^2 u}{\partial r^2} + \frac{1}{r} \frac{\partial u}{\partial r} \right), \qquad 0 \leqslant r \leqslant R, \, t \geqslant 0,$$

$$u(R, t) = 0, \qquad t \geqslant 0.$$

Here R is the radius of the membrane, $u(r, t)$ is the transverse displacement of a point distant r from the centre of the membrane at time t and a is a positive constant. Separate variables to obtain a singular Sturm–Liouville system. Find the eigenvalues of this system in terms of the zeros of the Bessel function J_0, and write down corresponding eigenfunctions.

9.3. Given the differential equation

$$f'' + P(x)f' + Q(x)f = g(x), \qquad a \leqslant x \leqslant b,$$

find a function p on $[a, b]$ such that the given equation is equivalent to

$$(pf')' + qf = pg$$

for a suitable function q on $[a, b]$. Assume P continuous.

9.4. Show that, if φ is an eigenfunction corresponding to an eigenvalue λ of the Sturm–Liouville system

$$(xf')' + \lambda xf = 0, \qquad 0 < x \leqslant R,$$
$$f \text{ bounded}, \ f(R) = 0,$$

then

$$\lambda \int_0^R x|\varphi(x)|^2 \, dx = \int_0^R x|\varphi'(x)|^2 \, dx.$$

Deduce that all eigenvalues of the system are positive.

9.5. The conduction of heat in a certain homogeneous circular cylinder of radius R and infinite length is governed by the equations

$$\frac{\partial u}{\partial t} = \frac{\partial^2 u}{\partial r^2} + \frac{1}{r}\frac{\partial u}{\partial r}, \qquad 0 \leqslant r \leqslant R, t \geqslant 0,$$

$$u(R, t) = 0, \qquad\qquad t \geqslant 0,$$

$$u(r, 0) = u_0(r), \qquad\quad 0 \leqslant r \leqslant R,$$

$$u(r, t) \text{ is bounded for } 0 \leqslant r \leqslant R, t \geqslant 0,$$

where u_0 is a non-zero real function with two continuous derivatives. Show that the normal modes of this system have the form

$$u(r, t) = cJ_0(\alpha r)\,e^{-\alpha^2 t}$$

where c is a constant, J_0 is the Bessel function of order 0 [i.e. $y = J_0(x)$ is the bounded solution of

$$xy'' + y' + xy = 0, \qquad y(0) = 1]$$

and αR is a zero of J_0.

9.6. Find the eigenvalues and eigenfunctions of the following Sturm–Liouville systems.

(a) $f'' + \lambda f = 0,$ $\qquad\qquad$ $f'(0) = 0 = f(1).$

(b) $f'' + \lambda f = 0,$ $\qquad\qquad$ $f(0) = 0 = 3f(1) + f'(1).$

(c) $f'' + f' + \lambda f = 0,$ $\qquad$ $f(0) = 0 = f(1).$

(d) $x^2 f'' + 3xf' + \lambda f = 0,$ $\quad$ $f(1) = 0 = f(e).$

Example (d) becomes a Sturm–Liouville problem on the substitution $x = e^t$. In (b) draw a graph to estimate the approximate value of the nth eigenvalue. Example (c) becomes a Sturm–Liouville problem on multiplication by an appropriate integrating factor.

9.7. Prove that the singular Sturm–Liouville system

$$(x^2 f')' + \lambda f = 0 \qquad \text{on } (0, 1],$$

$$f \text{ is bounded on } (0, 1], \ f(1) = 0,$$

has no eigenvalues.

[Find $\alpha \in \mathbb{C}$ such that $f(x) = x^\alpha$ is a solution of the differential equation, and obtain the general solution by putting $f = x^\alpha u$.]

10

Green's functions

We wish to diagonalize the self-adjoint linear operator L which arises in Sturm–Liouville theory, and by virtue of Chapter 8 we know how to diagonalize *compact* self-adjoint operators. Far from being compact, L is not even bounded. We shall nevertheless diagonalize L with the aid of the spectral theorem – by applying it to L^{-1} rather than L itself. To do this we need a representation of L^{-1}, so that we can see it is compact.

Consider the regular Sturm–Liouville problem RSL of the previous chapter (Definition 9.1) and let L and $D(L)$ be as defined above in (9.19). To find $L^{-1}g$, where $g \in L^2(a, b)$, is the same as to find $f \in D(L)$ such that $Lf = g$, or equivalently, to find f such that $f'' \in L^2(a, b)$,

$$\frac{d}{dx}\left(p\frac{df}{dx}\right) + qf = g, \tag{10.1}$$

$$\left.\begin{array}{l} \alpha f(a) + \alpha' f'(a) = 0, \\ \beta f(b) + \beta' f'(b) = 0. \end{array}\right\} \tag{10.2}$$

To see what kind of an answer we can hope for let us look at a simple special case:

$$\frac{d^2 f}{dx^2} = g, \\ f(0) = f(1) = 0. \tag{10.3}$$

Integrating the differential equation twice we get

$$\begin{aligned} f(x) &= \int_0^x d\xi \int_0^\xi g(\tau)\, d\tau + Ax + B \\ &= \int_0^x d\tau\, g(\tau) \int_\tau^x d\xi + Ax + B \\ &= \int_0^x (x - \tau)g(\tau)\, d\tau + Ax + B. \end{aligned}$$

Since $f(0) = 0$, $B = 0$, while the condition $f(1) = 0$ yields

$$A = -\int_0^1 (1 - \tau)g(\tau)\,d\tau.$$

Hence

$$f(x) = \int_0^x (x - \tau)g(\tau)\,d\tau - x\int_0^x + \int_x^1 (1 - \tau)g(\tau)\,d\tau$$

$$= \int_0^x \tau(x - 1)g(\tau)\,d\tau + \int_x^1 x(\tau - 1)g(\tau)\,d\tau.$$

Thus, for given g, we can write down the solution of the boundary value problem (10.3) as

$$f(x) = \int_0^1 k(x, \tau)g(\tau)\,d\tau,$$

where

$$k(x, \tau) = \begin{cases} \tau(x - 1) & \text{if } 0 \leqslant \tau < x \\ x(\tau - 1) & \text{if } x \leqslant \tau \leqslant 1. \end{cases}$$

There is a question as to what functions g are admissible here, since the indefinite integral of a discontinuous function need not be differentiable, but this solution is certainly valid whenever g is continuous on $[a, b]$.

We have obtained the solution of the special boundary value problem $Lf = g$ of (10.3) in the form $f = Kg$ where K is an integral operator. The kernel function $k(x, \tau)$ of this operator is called the *Green's function* for the boundary value problem. We shall show that the more general problem (10.1) also has a continuous Green's function, from which it will follow that L^{-1} is compact.

There is a standard technique for attacking inhomogeneous linear differential equations like (10.1): Lagrange's method of variation of parameters. Suppose we can find two linearly independent solutions u and v of the corresponding homogeneous equation

$$\frac{d}{dx}\left(p\frac{df}{dx}\right) + qf = 0. \tag{10.4}$$

We then look for a solution of (10.1) of the form

$$f = \varphi u + \psi v,$$

where φ, ψ are differentiable functions on $[a, b]$ to be chosen. We have

$$f' = \varphi u' + \psi v' + \varphi' u + \psi' v,$$

and if we choose φ, ψ to satisfy

$$\varphi' u + \psi' v = 0 \tag{10.5}$$

then

$$f' = \varphi u' + \psi v'. \tag{10.6}$$

Hence

$$(pf')' + qf = (p\varphi u')' + q\varphi u + (p\psi v')' + q\psi v$$
$$= \varphi'pu' + \varphi((pu')' + qu) + \psi'pv' + \psi((pv')' + qv)$$
$$= \varphi'pu' + \psi'pv'. \tag{10.7}$$

Thus, if φ, ψ are chosen to satisfy (10.5) then $f = \varphi u + \psi v$ will be a solution of (10.1) if and only if

$$p(\varphi'u' + \psi'v') = g. \tag{10.8}$$

Eliminating φ' between (10.5) and (10.8) we find that

$$\psi' \cdot p(uv' - vu') = ug. \tag{10.9}$$

Denote the left hand side of (10.1) by Mf, M being regarded as a differential operator whose domain is the set of $f \in L^2(a, b)$ such that f'' exists and belongs to $L^2(a, b)$ (M is given by the same formula as the operator L introduced earlier, but has a larger domain of definition). Since $Mu = 0 = Mv$, Lagrange's identity gives us

$$(p(uv' - vu'))' = uMv - vMu = 0.$$

Thus $p(uv' - vu')$ has a constant value, c say. Equation (10.9) becomes

$$c\psi' = ug. \tag{10.10}$$

Likewise, eliminating ψ' between (10.5) and (10.8) we obtain

$$c\varphi' = -vg. \tag{10.11}$$

Provided $c \neq 0$ we may define

$$\varphi(x) = c^{-1}\left(\int_x^b v(\tau)g(\tau)\,d\tau + A \right), \tag{10.12}$$

$$\psi(x) = c^{-1}\left(\int_a^x u(\tau)g(\tau)\,d\tau + B \right). \tag{10.13}$$

As long as g is continuous on $[a, b]$, φ and ψ will be differentiable and will satisfy (10.10) and (10.11), from which it follows that they satisfy (10.5) and (10.1) also. Let us summarize the result of these calculations.

10.1 Lemma   Suppose that u, v are solutions of the homogeneous differential equation (10.4) on $[a, b]$. If $g \in C[a, b]$ and the constant value c of $p(uv' - vu')$ is not zero, then the function

$$f = \varphi u + \psi v,$$

where φ, ψ are defined by (10.12), (10.13), is a solution of the inhomogeneous equation (10.1) for any values of the scalars A and B. $\square$

Subject to provisos, we have constructed a solution of the inhomogeneous equation (10.1) with two free scalar parameters. There is

every hope that we can choose these to satisfy the boundary conditions
(10.2). First we need to check that the homogeneous equation (10.4) does
have linearly independent solutions u and v. For this we invoke a standard
result about differential equations (see Simmons, 1972, p. 435).

10.2 Existence theorem   Let P, Q and R be continuous real functions on
an interval $[a, b]$. Let $x_0 \in [a, b]$ and let $y_0, y_1 \in \mathbb{R}$. Then the initial value
problem

$$\frac{d^2 y}{dx^2} + P(x)\frac{dy}{dx} + Q(x)y = R(x),$$

$$y(x_0) = y_0, \qquad y'(x_0) = y_1$$

has one and only one solution on the interval $[a, b]$, and this solution is
real.   □

In the case of the regular problem RSL, when $p > 0$ on $[a, b]$, the
differential equation (10.4) can be re-written

$$\frac{d^2 f}{dx^2} + \frac{p'}{p}\frac{df}{dx} + \frac{q}{p}f = 0, \tag{10.14}$$

so that the existence theorem applies. It follows that there exist non-zero
real solutions u, v of (10.4) such that

$$\alpha u(a) + \alpha' u'(a) = 0, \tag{10.15}$$

$$\beta v(b) + \beta' v'(b) = 0. \tag{10.16}$$

This is the most convenient choice for showing that the boundary
conditions (10.2) can be satisfied: we obtain a solution of the system (10.1),
(10.2) simply by taking both scalars A, B in Lemma 10.1 to be zero (subject
still to the proviso $c \neq 0$). That is, if

$$f = \varphi u + \psi v$$

where

$$\varphi(x) = c^{-1}\int_x^b v(\tau)g(\tau)\,d\tau, \tag{10.17}$$

$$\psi(x) = c^{-1}\int_a^x u(\tau)g(\tau)\,d\tau, \tag{10.18}$$

$a \leqslant x \leqslant b$, then f is a solution of the inhomogeneous system (10.1)–(10.2).
We have already seen that f satisfies the inhomogeneous differential
equation (10.1). To check the boundary conditions, note that $\psi(a) = 0$ and
recall from (10.6) that

$$f' = \varphi u' + \psi v'.$$

Thus
$$\alpha f(a) + \alpha' f'(a) = \varphi(a)(\alpha u(a) + \alpha' u'(a))$$
$$= 0.$$
Likewise f satisfies the boundary condition at b.

These calculations lead to the following result.

10.3 Theorem   Suppose that 0 is not an eigenvalue of the regular Sturm–Liouville system RSL. Then for any $g \in C[a, b]$ the corresponding inhomogeneous boundary value problem (10.1)–(10.2) has the unique solution

$$f(x) = \int_a^b k(x, \tau) g(\tau) \, d\tau$$

where

$$k(x, \tau) = \begin{cases} c^{-1} v(x) u(\tau) & \text{if } a \leqslant \tau \leqslant x \leqslant b \\ c^{-1} u(x) v(\tau) & \text{if } a \leqslant x < \tau \leqslant b \end{cases} \qquad (10.19)$$

where u, v are non-zero real solutions of the homogeneous equation (10.4) which satisfy (10.15), (10.16) respectively, and c is the non-zero constant value of $p(uv' - vu')$. Furthermore, f'' exists and is continuous on $[a, b]$.

We first establish a consequence of the eigenvalue assumption.

10.4 Lemma   Under the assumptions of Theorem 10.3, $uv' - vu'$ is non-zero at every point of $[a, b]$.

Proof. Suppose that

$$uv' - vu' = \det \begin{bmatrix} u & v \\ u' & v' \end{bmatrix}$$

vanishes at $x_0 \in [a, b]$. Then there exist scalars λ, μ, not both zero, such that

$$\lambda u(x_0) + \mu v(x_0) = 0,$$
$$\lambda u'(x_0) + \mu v'(x_0) = 0.$$

Thus the function $f = \lambda u + \mu v$ is a solution of the homogeneous equation (10.4) satisfying $f(x_0) = f'(x_0) = 0$. By the uniqueness assertion in Theorem 10.2, f must be the zero solution of (10.4). That is, u and v are linearly dependent, and each is a non-zero scalar multiple of the other. It follows from (10.16) that

$$\beta u(b) + \beta' u'(b) = 0.$$

Thus u is a non-zero solution of RSL with $\lambda = 0$, making 0 an eigenvalue of RSL, contrary to hypothesis. $\square$

Proof of Theorem 10.3. Since both p and $uv' - vu'$ are non-zero at every point of $[a, b]$, the constant value c of $p(uv' - vu')$ is not zero. Since 0 is not an eigenvalue of L, the equation $Lf = g$ certainly has at most one solution in $D(L)$. The foregoing calculations show that a solution of (10.1)–(10.2) is

$$f = \varphi u + \psi v$$

where φ, ψ are given by (10.12), (10.13). That is,

$$f(x) = u(x)c^{-1} \int_x^b v(\tau)g(\tau)\, d\tau + v(x)c^{-1} \int_a^x u(\tau)g(\tau)\, d\tau$$

$$= \int_a^b k(x,\tau)g(\tau)\, d\tau.$$

It remains to check that f has two continuous derivatives. Certainly u'' exists: this is part of the assertion that u is a solution of a second-order equation, and the relation

$$u'' = -\frac{p'u'}{p} - \frac{qu}{p}$$

(cf. (10.14)) shows that u'' is continuous. From (10.10) and (10.11) we see that φ', ψ' are continuous. Since $f' = \varphi u' + \psi v'$ and each function on the right hand side is continuously differentiable, f' is too.   □

Thus the Green's function for the inhomogeneous regular Sturm–Liouville system (10.1)–(10.2) is the function k defined by (10.19). Since both u and v are continuous, the functions $c^{-1}v(x)u(\tau)$ and $c^{-1}u(x)v(\tau)$ are continuous on the closed triangles $a \leqslant \tau \leqslant x \leqslant b$, $a \leqslant x \leqslant \tau \leqslant b$. Since the two functions agree on the diagonal where the two triangles meet, k is continuous on their union, the compact rectangle $[a,b] \times [a,b]$. Hence k is bounded: say

$$|k(x,\tau)| \leqslant M, \qquad a \leqslant x,\tau \leqslant b,$$

and so

$$\int_a^b \int_a^b |k(x,\tau)|^2\, dx\, d\tau \leqslant M(b-a)^2.$$

By Theorem 8.8, the integral operator K with kernel function k is Hilbert–Schmidt, and so it is compact, by Theorem 8.7. K is also Hermitian. By Example 7.14(ii), K^* is the integral operator with kernel function

$$k^*(x,\tau) = \overline{k(\tau, x)}.$$

Since u and v are real, one can see by inspecting (10.19) that $k^* = k$, and so $K^* = K$.

10.1   Compactness of the inverse of a Sturm–Liouville operator

The foregoing reasoning establishes the compactness property we need. To summarize:

10.5 Theorem   Suppose 0 is not an eigenvalue of the regular Sturm–Liouville problem RSL. Let k be as in Theorem 10.3 and let K be the integral operator on $L^2(a, b)$ with kernel k:

$$Kg(x) = \int_a^b k(x, \tau) g(\tau) \, d\tau.$$

Then K is a compact Hermitian operator and, for any $g \in C[a, b]$, the solution of the inhomogeneous boundary value problem

$$Lf = g, \qquad f \in D(L)$$

is

$$f = Kg. \qquad\qquad\qquad\qquad\qquad \Box$$

This theorem comes close to saying that $K = L^{-1}$. Two things are lacking, however: we have only shown the validity of the solution $f = Kg$ for continuous g, and we do not know that the range of L (and hence the domain of L^{-1}) is the whole of $L^2(a, b)$. In fact, the way we have defined $D(L)$, the range of L is a proper subspace of $L^2(a, b)$. It is possible to enlarge the domain of L by relaxing the sense in which f'' has to exist and so make L surjective onto $L^2(a, b)$. This modified L does indeed have K as its inverse, but the proper definitions and proof require some measure-theoretic technicalities which we shall avoid.

10.6 Lemma   Suppose 0 is not an eigenvalue of RSL, and let K be the integral operator described in Theorem 10.5. Then for any $g \in L^2(a, b)$, Kg is differentiable and

$$(Kg)' = \varphi u' + \psi v' \tag{10.20}$$

where φ, ψ are given by (10.17)–(10.18).

Proof. By definition of K,

$$Kg = \varphi u + \psi v.$$

Here the differentiability of u, v is guaranteed by Theorem 10.2. When g is continuous then φ and ψ are differentiable, by the fundamental theorem of calculus, so that Kg is differentiable and the desired formula for $(Kg)'$ follows from the product rule and the relations (10.10), (10.11). When g is a discontinuous L^2 function then φ and ψ will not in general be differentiable, but since the formula (10.20) does not involve φ' or ψ' there is a good chance it will still hold. Certainly the right hand side of (10.20) is a continuous function, so that (10.20) is equivalent to

$$Kg(x) = Kg(a) + \int_a^x (\varphi u' + \psi v')(\tau) \, d\tau$$

$$= u(a)c^{-1} \int_a^b v(\tau) g(\tau) \, d\tau + \int_a^x (\varphi u' + \psi v')(\tau) \, d\tau, \tag{10.21}$$

for all $x \in [a, b]$ (not just 'almost all'). Introduce three linear mappings

$$\tilde{K}, M, N : L^2(a, b) \to C[a, b]$$

by

$$\tilde{K}g = Kg,$$

$$Mg(x) = u(a)c^{-1} \int_a^b v(\tau)g(\tau)\,d\tau,$$

$$Ng(x) = \int_a^x (\varphi u' + \psi v')(\tau)\,d\tau,$$

where φ, ψ are given by (10.17), (10.18). It is straightforward to check that $\tilde{K}, M, N$ do map $L^2(a, b)$ into $C[a, b]$, and furthermore that they are continuous with respect to the Hilbert space norm on $L^2(a, b)$ and the supremum norm $\|\cdot\|_\infty$ on $C[a, b]$. Let us prove this for N. Since an indefinite integral of an L^2 function is continuous (Problem 10.1) φ and ψ are continuous. Thus $\varphi u' + \psi v'$ is continuous, and so its indefinite integral Ng is continuous on $[a, b]$. For any $x \in [a, b]$,

$$|Ng(x)| \leqslant \int_a^x |(\varphi u' + \psi v')(\tau)|\,d\tau$$

$$\leqslant (b - a)\{\|\varphi\|_\infty \|u'\|_\infty + \|\psi\|_\infty \|v'\|_\infty\}.$$

And

$$|\varphi(x)| = c^{-1}\left| \int_x^b v(\tau)g(\tau)\,d\tau \right|$$

$$\leqslant c^{-1} \int_a^b |v(\tau)||g(\tau)|\,d\tau$$

$$\leqslant c^{-1} \|v\|_{L^2} \|g\|_{L^2}.$$

Likewise

$$\|\psi\|_\infty \leqslant c^{-1} \|u\|_{L^2} \|g\|_{L^2}.$$

Hence

$$\|Ng\|_\infty \leqslant c^{-1}(b - a)\{\|v\|_{L^2} \|u'\|_\infty + \|u\|_{L^2} \|v'\|_\infty\} \|g\|_{L^2},$$

and so $N : L^2(a, b) \to C[a, b]$ is a bounded linear operator.

It follows that $\tilde{K} - M - N$ is a continuous mapping from $L^2(a, b)$ to $C[a, b]$; it is identically zero on the dense subspace $C[a, b]$ of $L^2(a, b)$, and hence is the zero mapping. That is, (10.21) is valid for all $g \in L^2(a, b)$, and so (10.20) is also. $\square$

10.7 Theorem Suppose that 0 is not an eigenvalue of RSL, and let K be the integral operator described in Theorem 10.5. Then

(i) 0 is not an eigenvalue of K, and

(ii) λ is an eigenvalue of L if and only if λ^{-1} is an eigenvalue of K.

Furthermore, the eigenvectors of L corresponding to λ coincide with the eigenvectors of K corresponding to λ^{-1}.

Proof. (i) Suppose $Kg = 0$. In the notation of Theorem 10.3,

$$\varphi u + \psi v = 0.$$

By Lemma 10.6,

$$\varphi u' + \psi v' = (Kg)' = 0.$$

Thus

$$\begin{bmatrix} u & v \\ u' & v' \end{bmatrix} \begin{bmatrix} \varphi \\ \psi \end{bmatrix} = 0.$$

By Lemma 10.4, the 2×2 matrix is non-singular at every point of $[a, b]$, and hence $\varphi = \psi = 0$ identically. Thus

$$\int_x^b v(\tau)g(\tau)\, d\tau = 0,$$

$$\int_a^x u(\tau)g(\tau)\, d\tau = 0$$

on $[a, b]$. Thus ug and vg are the zero function in $L^2(a, b)$ (strictly, they are zero almost everywhere). Again by Lemma 10.4, u and v never vanish simultaneously, and so $g = 0$. That is, 0 is not an eigenvalue of K.

Proof. (ii) Suppose that λ is an eigenvalue of L with corresponding eigenvector $f \in D(L)$:

$$Lf = \lambda f$$

and $f \neq 0, \lambda \neq 0$. Since $f \in D(L)$, λf is continuous, and so, by Theorem 10.3, $Lf = \lambda f$ is equivalent to

$$f = K(\lambda f),$$

and so

$$Kf = \lambda^{-1} f.$$

Thus λ^{-1} is an eigenvalue of K and f is a corresponding eigenvector.

Conversely, suppose that μ is an eigenvalue of K with eigenvector $g \in L^2(a, b)$: then $Kg = \mu g$ with $g \neq 0, \mu \neq 0$. Thus $g = \mu^{-1} Kg$, and Lemma 10.6 shows that g is continuous. By Theorem 10.3, $Lf = g$ has the solution $f = Kg$: thus

$$LKg = g.$$

which implies

$$L(\mu g) = g$$

and so

$$Lg = \mu^{-1} g$$

as desired. $\qquad\qquad\qquad\qquad\qquad\qquad\qquad\qquad\qquad\qquad\quad \square$

In a sense we have succeeded in inverting the Sturm–Liouville differential operator L. The foregoing shows that we can regard L as the inverse of a compact Hermitian operator K, and we even have a concrete representation of K as an integral operator whose kernel k is a reasonably accessible function. In the next chapter we shall claim the pay-off for this step by applying the spectral theorem to K and deriving consequences for the description of physical systems. For the moment let us learn some other lessons. Firstly, the length and difficulty of our calculations illustrate the fact that although the general theory of compact operators on Hilbert space is beautiful, powerful and important, it is only one of several ingredients that go into the analysis of a physical system. To get from a Sturm–Liouville system to a compact Hermitian operator we needed a standard technique (variation of parameters) and a fundamental result (Theorem 10.2) from the theory of differential equations, as well as quite a lot of ingenuity. And even so we have only dealt with regular Sturm–Liouville systems. Like most authors, I shall shirk the technicalities of extending the results to singular systems. Problem 9.7 indicates one difficulty. A thoroughgoing study of the singular case can be found in Chapter 13 of Dunford and Schwartz (1963). Another moral to be drawn is that, to go further with the theory, one would have to grasp the nettle of the Lebesgue integral. We have skirted round it, and though this was sufficient for the gaining of an understanding of one of the major roles of operator theory, it would not do if we wished to obtain the strongest results and most elegant statements.

10.2   Problems

10.1.   Show that an indefinite integral of an L^2 function is continuous. That is, if $f \in L^2(a, b)$ and $x_0 \in [a, b]$, then F defined by

$$F(x) = \int_{x_0}^{x} f(\tau)\, d\tau$$

is continuous on $[a, b]$. Show that F need not be differentiable on $[a, b]$ by taking f to be a step function.

10.2. Let ω be positive, not an integer multiple of π. Find the Green's function for the boundary value problem

$$f'' + \omega^2 f = g,$$
$$f'(0) = 0 = f'(1).$$

What happens if we try this with $\omega = 0$?

10.3. Find the Green's function for the inhomogeneous Sturm–Liouville problem

$$f'' = g,$$
$$f(0) = 0, \qquad f(1) + f'(1) = 0.$$

10.4. An alternative proof that 0 is not an eigenvalue of K, Theorem 10.7(i).

   (1) Show that the range of K contains the space

   $$E = \{ f \in D(L) : f'' \text{ is continuous on } [a, b] \}.$$

   (2) Show that E is dense in $L^2(a, b)$.

   (3) Deduce that $\operatorname{Ker} K = \{0\}$ using Problem 7.29.

This proof does without Lemma 10.6, and is the proof an operator theorist would give, as he would regard statement (2) as obvious. Starting from scratch, however, we should probably take longer to prove (2) than Lemma 10.6.

10.5. Show that the existence theorem 10.2 ceases to hold if the differential equation is replaced by

$$(py')' + qy = 0$$

where p is allowed to vanish at an end point of $[a, b]$ (use the solution of Problem 9.7).

10.6. Give an example to show that the differential operator

$$L : D(L) \to L^2(a, b)$$

corresponding to a regular Sturm–Liouville system, as defined in (9.19), need not be surjective.

11

Eigenfunction expansions

An eigenfunction expansion is a Fourier series expansion in an L^2 space with respect to a complete orthonormal sequence consisting of the eigenvectors of a differential operator. Such expansions arise directly from the idea of diagonalizing operators. Consider, for example, the Sturm–Liouville differential operator L of the previous two chapters. To diagonalize L is to find an orthonormal basis of $L^2(a, b)$ with respect to which L has a diagonal matrix – or, in other words, an orthonormal basis consisting of eigenvectors of L. Once such a basis is found, the action of L on a function f is very easy to describe in terms of the Fourier series of f with respect to this basis. In this chapter we shall show that Sturm–Liouville operators do indeed admit complete orthonormal sequences of eigenfunctions and examine the implications of this fact for the method of separation of variables.

We recall the regular Sturm–Liouville system

$$\text{RSL:} \quad (pf')' + (q + \lambda\rho)f = 0, \qquad a \leqslant x \leqslant b \tag{11.1}$$

$$\alpha f(a) + \alpha' f'(a) = 0, \tag{11.2}$$

$$\beta f(b) + \beta' f'(b) = 0.$$

Here p, q and ρ are continuous real-valued functions on $[a, b]$, p and ρ are strictly positive on $[a, b]$ and p is continuously differentiable on $[a, b]$.

We shall say that an eigenvalue λ of RSL is *simple* if any two eigenfunctions of RSL corresponding to λ are linearly dependent.

11.1 Sturm–Liouville theorem The system RSL has an infinite sequence $(\lambda_j)_1^\infty$ of eigenvalues. Each eigenvalue is real and simple, and $|\lambda_j| \to \infty$ as $j \to \infty$. If φ_j is an eigenfunction of RSL corresponding to λ_j then $(\rho^{1/2}\varphi_j)_1^\infty$ is a complete orthogonal system in $L^2(a, b)$.

Proof. We need only put together the main results of Chapters 8 and 10 and add a couple of refinements. Assume to begin with that 0 is not an eigenvalue of RSL. Let $D(L)$ and L be as in the previous two chapters, so that $D(L)$ consists of the twice-differentiable functions on $[a, b]$ satisfying (11.2) and Lf is the right hand side of (9.19). According to Definition 9.3 λ is an eigenvalue of RSL if $Lf = -\lambda\rho f$ for some non-zero $f \in D(L)$. We have to watch the minus sign here as there is a slight terminological clash with Chapters 8 and 10. Consider first the case that ρ is identically 1 on (a, b): then λ is an eigenvalue of RSL if and only if $-\lambda$ is an eigenvalue of L. By Theorems 10.5 and 10.7 there is a compact Hermitian operator K on $L^2(a, b)$ (to be thought of as L^{-1}) such that K and L have the same eigenfunctions, corresponding eigenvalues being reciprocals of each other. By the spectral theorem (more precisely, Corollary 8.16), we can choose a complete orthonormal sequence (φ_j) of eigenvectors of K, and the corresponding eigenvalues (μ_j) constitute a sequence of real numbers tending to zero. By Theorem 10.7 the μ_j are all non-zero, and (μ_j^{-1}) is the sequence of eigenvalues of L. The eigenvalues of RSL are thus the numbers $\lambda_j = -\mu_j^{-1}$, $1 \leqslant j < \infty$; clearly λ_j is real and $|\lambda_j| \to \infty$. Moreover φ_j is an eigenfunction of RSL corresponding to λ_j, and (φ_j) is a complete orthonormal sequence in $L^2(a, b)$.

Now consider general ρ, continuous and positive on $[a, b]$. Informally speaking, λ is an eigenvalue of RSL with eigenfunction φ

$$\Leftrightarrow L\varphi = -\lambda\rho\varphi \Leftrightarrow -\lambda^{-1}\varphi = L^{-1}\rho\varphi$$
$$\Leftrightarrow -\lambda^{-1}\rho^{1/2}\varphi = \rho^{1/2}L^{-1}\rho^{1/2}\varphi$$
$$\Leftrightarrow -\lambda^{-1}\rho^{1/2}\varphi = (\rho^{1/2}L^{-1}\rho^{1/2})\rho^{1/2}\varphi$$

$\Leftrightarrow -\lambda^{-1}$ is an eigenvalue of $\rho^{1/2}L^{-1}\rho^{1/2}$ with eigenfunction $\rho^{1/2}\varphi$. More precisely, let M denote the operator on $L^2(a, b)$ given by $Mf = \rho^{1/2}f$. Since $\rho^{1/2}$ is real, $M^* = M$ (cf. Example 7.14(i)). Thus

$$(MKM)^* = M^*K^*M^* = MKM,$$

and so MKM is a compact Hermitian operator. By the spectral theorem we can pick a complete orthonormal sequence (f_j) in $L^2(a, b)$ consisting of eigenvectors of MKM. Let μ_j be the eigenvalue of MKM corresponding to f_j and let $\varphi_j = \rho^{-1/2}f_j$. Since both M and K are injective, MKM is also, and so the μ_j are all non-zero. Since ρ is positive and continuous on $[a, b]$ it is bounded away from zero (i.e. $\rho(x) \geqslant c > 0$ for some c and all $x \in [a, b]$), so that $\rho^{-1/2}$ is bounded and so $\varphi_j(= \rho^{-1/2}f_j)$ belongs to $L^2(a, b)$. We must show that φ_j is an eigenfunction of RSL with corresponding eigenvalue $\lambda_j = -\mu_j^{-1}$. We have

$$MKMf_j = \mu_j f_j$$

which, in terms of φ_j, becomes

$$K(\rho\varphi_j) = \mu_j\varphi_j.$$

By Lemma 10.6, $K(\rho\varphi_j)$ is continuous, and so φ_j is also. Thus $\rho\varphi_j$ is continuous, and so, by Theorem 10.5, the equation

$$Lf = \rho\varphi_j$$

has the unique solution

$$f = K(\rho\varphi_j)$$

$$= \mu_j\varphi_j.$$

Hence

$$L(\mu_j\varphi_j) = \rho\varphi_j$$

and so

$$L\varphi_j = -(-\mu_j^{-1})\rho\varphi_j$$

$$= -\lambda_j\rho\varphi_j.$$

The λ_j are real, $|\lambda_j| \to \infty$ and $(\rho^{1/2}\varphi_j) = (f_j)$ is a complete orthonormal sequence.

We prove simplicity of the eigenvalues. Suppose there are linearly independent eigenfunctions f, g of RSL corresponding to some eigenvalue λ. Then for any scalars A, B, not both zero,

$$h = Af + Bg$$

is also an eigenfunction of RSL corresponding to λ. A little two-dimensional linear algebra (which will be left to the reader) shows that A and B can be chosen so that

$$h(a) = h'(a) = 0.$$

By the uniqueness assertion in Theorem 10.2, h must be the zero solution of

$$(L + \lambda\rho)h = 0,$$

contradicting linear independence of f and g. Thus every eigenvalue of RSL is simple.

We have proved all the conclusions of Theorem 11.1 subject to the hypothesis that 0 is not an eigenvalue of RSL. To remove this hypothesis pick a real number μ which is not an eigenvalue of RSL: this is possible by Theorem 9.10. Let RSL_μ be the Sturm–Liouville system obtained by replacing q by $q + \mu\rho$ in RSL. Then f is an eigenfunction of RSL_μ corresponding to the eigenvalue λ if and only if

$$(pf')' + (q + \mu\rho + \lambda\rho)f = 0,$$

which is possible if and only if $\lambda + \mu$ is an eigenvalue of RSL. Thus 0 is not an eigenvalue of RSL_μ, and so the conclusions of Theorem 11.1 hold for RSL_μ. It follows that they hold for RSL also. □

Theorem 11.1 is only the starting point of a major branch of analysis. More detailed information about the distribution of eigenvalues is needed for an understanding of physical systems, and a great deal of classical analysis has gone into providing it. Here, as a sample, are three statements valid for the general regular Sturm–Liouville system:

$\lambda_j \to \infty$ as $j \to \infty$ (so that only finitely many λ_j are negative);

$$\sum_{\lambda_j \neq 0} \frac{1}{\lambda_j} \text{ converges;}$$

φ_j has exactly j zeros on $[a, b]$.

There are also elaborate estimates of the number of eigenvalues in an interval, and occasionally asymptotic formulae for λ_j. The results are significant in particular for quantum physics. A classic study of such questions is Titchmarsh (1962). An important question for applications is whether an eigenfunction expansion converges in some stronger sense than in the L^2-norm, for example uniformly on $[a, b]$, and there are results analogous to the corresponding ones for classical Fourier series. Another major direction of research is the extension of Theorem 11.1 to singular Sturm–Liouville systems. Singular systems need not have any eigenvalues at all (see Problem 9.7) and so the theory of eigenfunction expansions will certainly have to be modified if it is to tell us anything about such systems. Discussions can be found in Dunford and Schwartz (1963, Chapter 13) and Tricomi (1961, Chapter 9).

11.1   Solution of the hanging chain problem

What use is it to know that a Sturm–Liouville system has a complete sequence of eigenfunctions? In principle it enables us to write down the solution of the initial boundary value problem from which the Sturm–Liouville system arose. Let us illustrate this with the hanging chain example of Chapter 9. Recall the problem

$$\text{HC:} \quad \frac{\partial^2 u}{\partial t^2} = \frac{\partial}{\partial x}\left(x \frac{\partial u}{\partial x} \right),$$

$$u(L, t) = 0, \qquad 0 \leqslant t < \infty,$$

$$u(x, 0) = u_0(x), \qquad 0 \leqslant x \leqslant L,$$

$$\frac{\partial u}{\partial t}(x, 0) = 0, \qquad 0 \leqslant x \leqslant L,$$

$$\sup|u(x, t)| < \infty, \qquad 0 \leqslant x \leqslant L, 0 \leqslant t < \infty.$$

Our aim is to write down the solution $u(x, t)$ of this system in terms of the eigenfunctions of the singular Sturm–Liouville system

$$\text{HCSL:} \quad \frac{d}{dx}\left(x\frac{df}{dx}\right) + \lambda f = 0, \qquad 0 < x \leqslant L,$$

$$f \text{ is bounded on } (0, L], \qquad f(L) = 0.$$

11.2 Lemma   All the eigenvalues of HCSL are positive.
Proof. Let φ be a unit eigenfunction corresponding to λ, and let L be the differential operator corresponding to HCSL. Then

$$\lambda = -(L\varphi, \varphi)$$

$$= -\int_0^L (x\varphi')'\bar{\varphi}\, dx$$

$$= -[x\varphi'\bar{\varphi}]_0^L + \int_0^L x|\varphi'(x)|^2\, dx > 0. \qquad \square$$

Denote the eigenvalues of HCSL by $(\lambda_j)_1^\infty$ and let φ_j be a unit eigenfunction corresponding to λ_j. As in the discussion in Chapter 9, we are hoping to express u as an infinite sum of normal modes, which means that

$$u(x, t) = \sum_1^\infty c_j\varphi_j(x)\cos \omega_j t$$

where $\omega_j = \sqrt{\lambda_j}$. If this is true for all t then, putting $t = 0$, we have

$$u_0(x) = \sum_1^\infty c_j\varphi_j(x), \tag{11.3}$$

which is to say that the c_j are the Fourier coefficients of u_0 with respect to the sequence of eigenfunctions of HCSL. The relation (11.3) can only be satisfied if the φ_js form a *complete* orthonormal sequence in $L^2(0, L)$. Unfortunately, but typically, the elegant general theory we have derived just fails to give us the information we want here – it applies to regular systems, while HCSL is a singular system. Quite a lot more work is needed to establish the completeness of the φ_j. This work has been done (see Watson, 1944), so we shall assume the result and proceed. The example is still valuable for showing the purpose of completeness results for eigenfunctions.

We expect, then, that the solution of HC is

$$u(x, t) = \sum_1^\infty c_j\varphi_j(x)\cos \omega_j t \tag{11.4}$$

where $\omega_j = \sqrt{\lambda_j}$ and $c_j = (u_0, \varphi_j)$. If term-by-term differentiation of (11.4) is valid we shall have

$$\frac{\partial}{\partial x}\left(x\frac{\partial u}{\partial x}\right) = \sum_1^\infty c_j \frac{d}{dx}\left(x\frac{d\varphi_j}{dx}\right)\cos\omega_j t$$

$$= -\sum_1^\infty c_j \lambda_j \varphi_j(x)\cos\omega_j t,$$

and a like process gives the same series for $\partial^2 u/\partial t^2$. However, it is far from clear that the right hand side makes sense, particularly since $\lambda_j \to \infty$ as $j \to \infty$. Nevertheless, the solution (11.4) can be justified.

11.3 Theorem Let u_0 be twice continuously differentiable on $[0, L]$. There is at most one solution $u(x, t)$ of HC which is twice continuously differentiable on $[0, L] \times [0, \infty)$. If there is a solution it is given by

$$u(\cdot, t) = \sum_1^\infty c_j(\cos\omega_j t)\varphi_j \tag{11.5}$$

where $(\lambda_j)_1^\infty$ is the sequence of eigenvalues of the singular Sturm–Liouville system HCSL, φ_j is a unit eigenfunction corresponding to λ_j, $\omega_j = \sqrt{\lambda_j}$ and $c_j = (u_0, \varphi_j)$. The expansion (11.5) holds with respect to the norm of $L^2(0, L)$ for every $t \geqslant 0$.

Proof. Let u be a solution of HC. For each $t \geqslant 0$ let

$$c_j(t) = (u(\cdot, t), \varphi_j).$$

Here and in the statement of the theorem, $u(\cdot, t)$ denotes the function on $[0, L]$ whose value at x is $u(x, t)$. As asserted above, the φ_js form a complete orthonormal sequence in $L^2(0, L)$, and hence, for all $t \geqslant 0$,

$$u(\cdot, t) = \sum_1^\infty c_j(t)\varphi_j \tag{11.6}$$

in $L^2(0, L)$. We show next that $c_j(\cdot)$ is twice differentiable on $[0, \infty)$ and that

$$\ddot{c}_j(t) = \left(\frac{\partial^2 u}{\partial t^2}(\cdot, t), \varphi_j\right) \tag{11.7}$$

(dots denote derivatives with respect to t). For $t \geqslant 0$ and $h \geqslant -t$ we have

$$\frac{c_j(t+h) - c_j(t)}{h} - \left(\frac{\partial u}{\partial t}(\cdot, t), \varphi_j\right)$$

$$= \left(\frac{u(\cdot, t+h) - u(\cdot, t)}{h} - \frac{\partial u}{\partial t}(\cdot, t), \varphi_j\right)$$

$$= \int_0^L \left\{\frac{u(x, t+h) - u(x, t)}{h} - \frac{\partial u}{\partial t}(x, t)\right\}\overline{\varphi_j(x)}\,dx. \tag{11.8}$$

For fixed t the integrand tends to zero as $h \to 0$ at every point of $[0, L]$, simply by the definition of $\partial u/\partial t$. At this point we need a property of the

Lebesgue integral: the dominated convergence theorem asserts that the integral itself tends to zero as $h \to 0$ provided that there exists a function g which has a finite integral over $[0, L]$ and has modulus no less than the modulus of the integrand at each point of $[0, L]$. We can find such a function g as follows. With t fixed, let M be the supremum of

$$\left| \frac{\partial u}{\partial t}(x, t') \right| \qquad \text{for } 0 \leqslant x \leqslant L, 0 \leqslant t' \leqslant t + 1.$$

Since $\partial u / \partial t$ is continuous, by assumption, and $[0, L] \times [0, t+1]$ is compact, M is finite. Let

$$g(x) = 2M \left| \varphi_j(x) \right|.$$

By the mean value theorem, for some $\theta \in (0, 1)$,

$$\left| \frac{u(x, t+h) - u(x, t)}{h} \right| = \left| \frac{\partial u}{\partial t}(x, t + \theta h) \right|$$

$$\leqslant M,$$

provided $-t \leqslant h \leqslant 1$. It follows that, for these values of h, the integrand in (11.8) has modulus at most $g(x)$. Since φ_j is continuous on $[0, L]$ it is bounded and so has a finite integral over $[0, L]$. Thus the dominated convergence theorem does apply, and the right hand side of (11.8) tends to zero as $h \to 0$. That is to say, $c_j(\cdot)$ is differentiable at t and

$$\dot{c}_j(t) = \left(\frac{\partial u}{\partial t}(\cdot, t), \varphi_j \right). \tag{11.9}$$

A repetition of the same argument, with the slightest of variations, shows that $\dot{c}_j(\cdot)$ is also differentiable and that $\ddot{c}_j$ is given by (11.7).

Now we show that $c_j(\cdot)$ satisfies the equation of simple harmonic motion. Since u is a solution of HC, (11.7) implies that

$$\ddot{c}_j(t) = \left(\frac{\partial}{\partial x} \left(x \frac{\partial u}{\partial x} \right) (\cdot, t), \varphi_j \right)$$

$$= (Lu(\cdot, t), \varphi_j),$$

where L is the Sturm–Liouville operator corresponding to the system HCSL (there is little danger of confusion between the two meanings of L). By self-adjointness of L,

$$\ddot{c}_j(t) = (u(\cdot, t), L\varphi_j)$$

$$= (u(\cdot, t), -\lambda_j \varphi_j)$$

$$= -\lambda_j c_j(t).$$

This is the equation of simple harmonic motion. It has general solution

$$c_j(t) = A_j \cos \omega_j t + B_j \sin \omega_j t$$

where A_j, B_j are arbitrary constants. Since $\partial u/\partial t(\,\cdot\,,0) = 0$, (11.9) gives

$$\dot{c}_j(0) = \left(\frac{\partial u}{\partial t}(\,\cdot\,,0), \varphi_j\right) = (0, \varphi_j) = 0.$$

It follows that $B_j = 0$. Furthermore

$$A_j = c_j(0) = (u(\,\cdot\,,0), \varphi_j)$$
$$= (u_0, \varphi_j) = c_j.$$

Hence

$$c_j(t) = c_j \cos \omega_j t$$

and so, by (11.6)

$$u(\,\cdot\,,t) = \sum_1^\infty c_j(\cos \omega_j t)\varphi_j$$

in $L^2(0, L)$.   □

The conditional nature of Theorem 11.3 would not trouble a physicist or engineer: the physical origin of the system appears to guarantee that there will be at least one solution. The applied mathematician would be less willing to take on trust that there is a solution which is twice continuously differentiable, and would turn to a standard work on the theory of linear partial differential equations, such as Friedman (1969), to discover what conditions on u_0 will ensure such a solution.

If an engineering company were manufacturing a machine which had one component which was a uniform flexible chain subject to small oscillations in a plane then it could at the design stage use the above solution to analyse how the completed machine would perform. Typical initial displacements u_0 would be expanded in eigenfunctions and the subsequent motion should be described approximately by (11.5) (see also Problem 11.2). This is perhaps not very realistic, but of course similar methods work for more complicated examples. Manufacturers of air compressors may have to use expansions in Bessel functions to calculate the transfer of heat in their devices. In the past this involved long hand calculation and could hold up other design work for weeks. Fortunately this form of intellectual drudgery is no longer necessary and the calculations are performed in minutes on computers.

11.2   Problems

11.1.   By applying the Sturm–Liouville theorem to the system

$$f'' + \lambda f = 0,$$
$$f(0) = 0 = f'(\pi),$$

show that, for any $g \in L^2(0, \pi)$,

$$g(x) = \sum_{j=1}^{\infty} C_j \sin(j - \tfrac{1}{2})x$$

in the norm of $L^2(0, \pi)$, where

$$C_j = \frac{2}{\pi} \int_0^{\pi} g(x) \sin(j - \tfrac{1}{2})x \, dx, \qquad j \in \mathbb{N}.$$

11.2. Make the solution of the hanging chain problem given in Theorem 11.3 explicit by using the expressions for ω_j and φ_j in terms of J_0 and its zeros which were given in Chapter 9.

$$[\omega_j = z_j/2\sqrt{L}, \qquad \varphi_j(x) = J_0(2\omega_j\sqrt{x})/M_j$$

where $M_j = \|J_0(2\omega_j\sqrt{x})\|_{L^2(0, L)}$. Substitute in (11.5) to obtain

$$u(x, t) = \sum_1^{\infty} C_j \cos(\omega_j t) J_0(2\omega_j\sqrt{x})$$

where

$$C_j = (u_0, \varphi_j)/M_j$$
$$= \frac{\int_0^L u_0(x) J_0(2\omega_j\sqrt{x}) \, dx}{\int_0^L J_0(2\omega_j\sqrt{x})^2 \, dx}.]$$

11.3. Show that if the initial boundary value problem given in Problem 9.5 has a twice continuously differentiable solution u then this solution is given by

$$u(r, t) = \sum_{j=1}^{\infty} C_j J_0(\alpha_j r) e^{-\alpha_j^2 t} \tag{11.10}$$

where $\alpha_j = z_j/R$, $\{z_j : j \in \mathbb{N}\}$ are the zeros of J_0 and

$$C_j = \frac{(r^{1/2}u_0(r), r^{1/2}J_0(\alpha_j r))}{\|r^{1/2}J_0(\alpha_j r)\|^2}$$
$$= \frac{\int_0^R r u_0(r) J_0(\alpha_j r) \, dr}{\int_0^R r J_0(\alpha_j r)^2 \, dr}.$$

[Assume that the conclusions of the Sturm–Liouville theorem hold, even though the system in question is singular. Mimic the proof of Theorem 11.3, considering the Fourier expansion of $r^{1/2}u(r, t)$, for fixed t, with respect to the orthonormal basis $r^{1/2}J_0(\alpha_j r)$, $j \in \mathbb{N}$, in $L^2(0, R)$. (11.10) converges with respect to the norm

$$\|f\|^2 = \int_0^R |f(r)|^2 r \, dr.]$$

12

Positive operators and contractions

The origins and early successes of Hilbert space methods were in the context of differential and integral equations, and in the last three chapters we have seen an illustration of their effectiveness. Analysts quickly recognized the beauty and power of the new ideas, and set about applying them and extending them to the other linear operators which abound in analysis. This process is still continuing: most recently it is composition operators which have suddenly begun to undergo intensive study. It would be premature to offer any account of these latest investigations here. Instead the rest of the book will be devoted to some applications of Hilbert space theory to that other great branch of analysis, complex function theory. The study of operators related to analytic functions began soon after the foundations of functional analysis were laid and has led to a rich theory which is still an active area of research: see Nikolskii (1985) for an up-to-date and penetrating monograph. As a sample of this great body of knowledge we shall study an aspect which has recently become topical in some questions of engineering design: it is always gratifying when pure mathematical constructs prove their worth outside the mathematical domain. An outline of the way operator theory makes contact with engineering will follow later. For the moment let us pursue the basics a little further. The notion of positive definite matrix is familiar from linear algebra: it has a natural generalization to infinite dimensions.

12.1 *Definitions* A bounded linear operator A on an inner product space H is *positive* if $(Ax, x) \geqslant 0$ for all $x \in H$.

The notation $A \geqslant 0$ will mean that A is positive. $A \geqslant B$ is defined to mean that $A - B \geqslant 0$; equivalently

$$A \geqslant B \Leftrightarrow (Ax, x) \geqslant (Bx, x) \qquad \text{for all } x \in H. \qquad \square$$

12.2 Examples (i) The zero and identity operators are positive.

(ii) The operator on ℓ^2 with matrix $\text{diag}\{\lambda_1, \lambda_2, \ldots\}$ is positive if and only if all the λ_ns are non-negative.

(iii) The operator M on $L^2(0, 1)$ defined by

$$(Mx)(t) = tx(t), \qquad 0 < t < 1,$$

is positive, since

$$(Mx, x) = \int_0^1 t|x(t)|^2 \, dt \geqslant 0$$

for any $x \in L^2(0, 1)$.

(iv) The operator on $\mathbb{C}^2$ with matrix $\begin{bmatrix} 1 & 2 \\ 2 & 1 \end{bmatrix}$ is not positive. We can see this using the determinantal criterion from linear algebra, but we can also see it directly by applying the matrix to the vector $(1, -1)$.

(v) If $T \in \mathcal{L}(H, K)$ for any pair of Hilbert spaces H, K then T^*T is a positive operator on H since

$$(T^*Tx, x) = (Tx, Tx) = \|Tx\|^2 \geqslant 0. \qquad \square$$

12.3 Exercise Prove that positive operators are Hermitian. Prove also that if M, N and P are bounded operators on a Hilbert space H and $M \geqslant N$ then $P^*MP \geqslant P^*NP$. $\square$

This involves expanding

$$(A(x + \lambda y), x + \lambda y) \geqslant 0$$

and doing some juggling with complex numbers, or else using the method of Problem 12.3.

Positive operators which have a diagonal matrix with respect to some orthonormal basis constitute quite a wide class: by the spectral theorem, they include all positive operators of the form $\lambda I + K$ with $\lambda \in \mathbb{R}$ and K compact. For diagonal positive operators the existence of square roots is obvious (recall Exercise 8.17). For other positive operators it is no longer obvious, but it is still true.

12.4 Theorem Let A be a positive operator on a Hilbert space H. There exists a unique positive square root $A^{1/2}$ of A; that is, there exists a unique positive operator B on H such that $B^2 = A$. Moreover there exists a sequence (p_n) of polynomials such that

$$p_n(A)x \to A^{1/2}x \qquad \text{as } n \to \infty$$

whenever $x \in H$ and A is a positive operator on H such that $\|A\| \leqslant 1$. $\square$

For the application in Chapter 15 it would be possible to make do with the case $\lambda I + K$ described above. However, it is more elegant to develop a general theory, and so it suits us to have this fundamental property of positive operators at our disposal.

A proof will be given in the Appendix.

12.5 Definition   Let H, K be Hilbert spaces. We say that $T \in \mathscr{L}(H, K)$ is a *contraction* if $\|T\| \leqslant 1$, a *strict contraction* if $\|T\| < 1$.   □

12.6 Theorem $T \in \mathscr{L}(H, K)$ is a contraction if and only if
$$I - T^*T \geqslant 0.$$
Proof. T is a contraction $\Leftrightarrow \|Tx\|^2 \leqslant \|x\|^2$, all $x \in H \Leftrightarrow (T^*Tx, x) \leqslant (x, x)$, all $x \in H \Leftrightarrow ((I - T^*T)x, x) \geqslant 0$, all $x \in H$.

Note further that T is a contraction $\Leftrightarrow T^*$ is a contraction $\Leftrightarrow I - TT^* \geqslant 0$.

12.7 Exercise   Let $A \in \mathscr{L}(G, H)$, $B \in \mathscr{L}(G, K)$ where G, H and K are Hilbert spaces and B is invertible. Show that AB^{-1} is a contraction if and only if $A^*A \leqslant B^*B$.   □

It is a remarkably useful fact (sometimes called 'Douglas' lemma') that the above statement remains very nearly true even when B is not invertible. Of course we cannot then write AB^{-1}, but we can speak of a solution Z of the equation $A = ZB$.

12.8 Theorem   Let $A \in \mathscr{L}(G, H)$, $B \in \mathscr{L}(G, K)$ where G, H and K are Hilbert spaces. There exists a contraction $Z \in \mathscr{L}(K, H)$ such that $A = ZB$ if and only if $A^*A \leqslant B^*B$.

Proof. ($\Rightarrow$) Suppose there is such a Z. Then $I - Z^*Z \geqslant 0$, and hence $B^*(I - Z^*Z)B \geqslant 0$. That is, $B^*B - A^*A \geqslant 0$.

($\Leftarrow$) Suppose $A^*A \leqslant B^*B$. Define Z on the range of B as follows. If $x \in \text{Range } B$ then $x = By$ for some $y \in G$: let $Zx = Ay$. This defines Zx unambiguously, for if $y' \in G$ is another vector such that $By' = x$ then $B(y - y') = 0$, and so
$$\|A(y - y')\|^2 = (A^*A(y - y'), y - y')$$
$$\leqslant (B^*B(y - y'), y - y')$$
$$= \|B(y - y')\|^2 = 0.$$
Thus $Ay = Ay'$.

Z so defined is a contraction from Range B to H, for if x, y are as above then

$$\|Zx\|^2 = \|Ay\|^2 = (A^*Ay, y)$$
$$\leqslant (B^*By, y) = \|By\|^2 = \|x\|^2.$$

We shall extend Z to the closure K_0 of Range B in K by continuity. Consider an arbitrary $x \in K_0$: there is a sequence (x_n) in Range B converging to x. Then (x_n) is a Cauchy sequence, and since Z is a contraction on Range B it follows that (Zx_n) is Cauchy in H. Hence (Zx_n) converges to some element $z \in H$, and we define Zx to be z. Again we must check that this defines Zx unambiguously. If (x'_n) is a second sequence in Range B converging to x then $x_n - x'_n \to 0$, and since Z is bounded, $Zx_n - Zx'_n \to 0$. Thus also $Zx'_n \to z$, and so we will always get the same value z whatever sequence (x_n) in Range B, converging to x, is chosen. Moreover,

$$\|Zx\| = \lim_n \|Zx_n\| \leqslant \lim_n \|x_n\| = \|x\|,$$

and so Z is a contraction on K_0. We now extend Z to a contraction on all of K. We have (Theorem 4.24)

$$K = K_0 \oplus K_0^\perp.$$

For $x \in K$ write $x = y + z$ with $y \in K_0$, $z \perp K_0$, and define Zx to be Zy. Then Z is a contraction from K to H and, by construction, $ZB = A$. $\square$

12.9 Corollary   For given operators A, B with the same codomain, there exists a contraction Z such that $A = BZ$ if and only if $AA^* \leqslant BB^*$.
Proof. Apply Theorem 12.8 to A^*, B^*. $\square$

12.1   Operator matrices

In Definition 4.26 we learnt what it means for a Hilbert space to be the orthogonal direct sum of two of its subspaces. There is another notion of direct sum, applying to a pair of spaces which are not necessarily sitting inside a given larger space: we can create such a larger space out of them.

12.10 Definition   Let H_1, H_2 be Hilbert spaces. The *orthogonal direct sum* $H_1 \oplus H_2$ of H_1, H_2 consists of the vector space direct sum of H_1 and H_2 together with the inner product

$$((x_1, x_2), (y_1, y_2)) = (x_1, y_1) + (x_2, y_2),$$

where $x_i, y_i \in H_i$, $i = 1, 2$. $\square$

As is customary we identify H_1, H_2 with the subspaces $H_1 \oplus \{0\}$, $\{0\} \oplus H_2$ of $H_1 \oplus H_2$. H_1, H_2 are then orthogonal subspaces of $H_1 \oplus H_2$, and $H_1 \oplus H_2$ is the orthogonal direct sum of H_1 and H_2 in the sense of Definition 4.26.

When thinking of operators on direct sums it is often helpful to think of elements of $H_1 \oplus H_2$ as 'column vectors' $\begin{bmatrix} x_1 \\ x_2 \end{bmatrix}$ with $x_i \in H_i$.

12.11 Definition   Let $A_{ij} \in \mathscr{L}(H_j, K_i)$, $i, j = 1, 2$. The *operator matrix*

$$A = \begin{bmatrix} A_{11} & A_{12} \\ A_{21} & A_{22} \end{bmatrix}$$

is the operator from $H_1 \oplus H_2$ to $K_1 \oplus K_2$ defined by

$$A \begin{bmatrix} x_1 \\ x_2 \end{bmatrix} = \begin{bmatrix} A_{11}x_1 + A_{12}x_2 \\ A_{21}x_1 + A_{22}x_2 \end{bmatrix},$$

$x_i \in H_i$, $i = 1, 2$. Operator matrices

$$[A_1 \quad A_2] : H_1 \oplus H_2 \rightarrow K,$$

$$\begin{bmatrix} A_1 \\ A_2 \end{bmatrix} : H \rightarrow K_1 \oplus K_2$$

are defined analogously.   □

12.12 Exercise   State and verify the rule for operator matrix multiplication. Show that

$$\begin{bmatrix} A_{11} & A_{12} \\ A_{21} & A_{22} \end{bmatrix}^* = \begin{bmatrix} A_{11}^* & A_{21}^* \\ A_{12}^* & A_{22}^* \end{bmatrix}.$$ □

12.13 Theorem   Let H, K_1 and K_2 be Hilbert spaces and let $A_i \in \mathscr{L}(H, K_i)$, $i = 1, 2$, be a contraction. $\begin{bmatrix} A_1 \\ A_2 \end{bmatrix}$ is a contraction from H to $K_1 \oplus K_2$ if and only if there exists a contraction $Z : H \rightarrow K_1$ such that
$$A_1 = Z(I - A_2^* A_2)^{1/2}.$$

Proof. By Theorem 12.6,

$$\begin{bmatrix} A_1 \\ A_2 \end{bmatrix} \text{ is a contraction} \Leftrightarrow I_H - [A_1^* \quad A_2^*]\begin{bmatrix} A_1 \\ A_2 \end{bmatrix} \geqslant 0$$

$$\Leftrightarrow I_H - A_1^* A_1^* - A_2^* A_2 \geqslant 0$$

$$\Leftrightarrow A_1^* A_1 \leqslant I - A_2^* A_2.$$

By Theorem 12.8 with $A = A_1$, $B = (I - A_2^* A_2)^{1/2}$, the last statement holds if and only if there exists a contraction Z such that $A_1 = Z(I - A_2^* A_2)^{1/2}$.   □

12.14 Corollary $[A_1 \quad A_2]$ is a contraction if and only if there exists a contraction Z such that $A_1 = (I - A_2 A_2^*)^{1/2} Z$.
Proof. Apply Theorem 12.13 to $[A_1 \quad A_2]^*$.   □

12.2   Möbius transformations

In the study of analytic functions in the unit disc $\mathbb{D}$ there is often occasion to use transformations of the form

$$\varphi(z) = \frac{z - \alpha}{1 - \bar{\alpha}z}$$

where $|\alpha| < 1$. The relation

$$1 - |\varphi(z)|^2 = \frac{(1 - |\alpha|^2)(1 - |z|^2)}{|1 - \bar{\alpha}z|^2}$$

shows that φ maps $\mathbb{D}$ to itself, and the formula

$$\varphi^{-1}(z) = \frac{z + \alpha}{1 + \bar{\alpha}z}$$

shows that φ is bijective on $\mathbb{D}$ and is analytic in both directions. For some purposes the closed unit ball of $\mathscr{L}(H, K)$ – that is, the set of contractions from H to K – can usefully be regarded as a generalization of the closed unit disc, and the analogues of φ play a significant role. We can also write

$$\varphi(z) = -\alpha + \frac{(1 - \alpha\bar{\alpha})z}{1 - \bar{\alpha}z}.$$

We wish to generalize this to an operator function of an operator argument which preserves the closed unit ball. In the absence of commutativity the right generalization is not obvious. It turns out to be as follows.

12.15 *Definition*   Let H, K be Hilbert spaces and let

$$T = \begin{bmatrix} A & B \\ C & D \end{bmatrix} : K \oplus H \to K \oplus H$$

be a unitary operator (this means that $T^*T = 1\,T^* = I_{K \oplus H}$). The transformation

$$\Psi_T(X) = B - AX(I + CX)^{-1}D$$

is called a *Möbius transformation*. It maps

$$\{X \in \mathscr{L}(H, K) : (I + CX)^{-1} \in \mathscr{L}(H)\}$$

into $\mathscr{L}(H, K)$. $\square$

The domain of Ψ_T contains all strict contractions. Since T is unitary, $\|T\| = 1$, and it is easy to show that $\|C\| \leqslant 1$. Thus, if $\|X\| < 1$,

$$\|CX\| \leqslant \|C\|\|X\| < 1,$$

and so $I + CX$ is invertible, by Theorem 7.10. What if $\|X\| = 1$? If it happens that $\|C\| < 1$ then we shall still have $\|CX\| < 1$ and so $\Psi_T(X)$ will be defined. In general, though, $\Psi_T(X)$ need not be defined when $\|X\| = 1$. Now let us prove the invariance of the set of contractions.

12.16 Theorem Let T be a unitary operator on $K \oplus H$. If $X \in$ domain Ψ_T then

$$I - \Psi_T(X)^* \Psi_T(X) = D^*(I + X^*C^*)^{-1}(I - X^*X)(I + CX)^{-1}D.$$

Proof. Since T is unitary,

$$\begin{bmatrix} A^* & C^* \\ B^* & D^* \end{bmatrix}\begin{bmatrix} A & B \\ C & D \end{bmatrix} = \begin{bmatrix} I & 0 \\ 0 & I \end{bmatrix};$$

that is,

$$A^*A + C^*C = I,$$

$$B^*B + D^*D = I,$$

$$A^*B + C^*D = 0,$$

$$B^*A + D^*C = 0.$$

Hence

$$I - \Psi_T(X)^* \Psi_T(X)$$

$$= I - \{B^* - D^*(I + X^*C^*)^{-1}X^*A^*\}\{B - AX(I + CX)^{-1}D\}$$

$$= I - B^*B + B^*AX(I + CX)^{-1}D + D^*(I + X^*C^*)^{-1}X^*A^*B$$

$$\quad - D^*(I + X^*C^*)^{-1}X^*A^*AX(I + CX)^{-1}D$$

$$= D^*D - D^*CX(I + CX)^{-1}D - D^*(I + X^*C^*)^{-1}X^*C^*D$$

$$\quad - D^*(I + X^*C^*)^{-1}X^*A^*AX(I + CX)^{-1}D$$

$$= D^*(I + X^*C^*)^{-1}\{(I + X^*C^*)(I + CX) - (I + X^*C^*)CX$$

$$\quad - X^*C^*(I + CX) - X^*A^*AX\}(I + CX)^{-1}D$$

$$= D^*(I + X^*C^*)^{-1}\{I - X^*C^*CX - X^*A^*AX\}(I + CX)^{-1}D$$

$$= D^*(I + X^*C^*)^{-1}(I - X^*X)(I + CX)^{-1}D. \qquad \square$$

Do you find this calculation delightful or repugnant? Both responses are to be found among professional mathematicians, though most analysts (including me) would think it beautiful. It is an unexpected triumph over non-commutativity. It is essentially due to C. L. Siegel and was discovered in the 1940s.

12.17 Corollary If $X \in$ domain Ψ_T then

$$I - \Psi_T(X)\Psi_T(X)^* = A(I + XC)^{-1}(I - XX^*)(I + C^*X^*)^{-1}A^*.$$

Proof. Either deduce it from Theorem 12.16 by the method proposed in Problem 12.14 or carry out the same kind of calculation as in the proof of Theorem 12.16. $\qquad \square$

12.18 Corollary If X is a contraction and $X \in$ domain Ψ_T then $\Psi_T(X)$ is a contraction.

Proof. Let $P = (I + CX)^{-1}D$. Since X is a contraction, $I - X^*X \geqslant 0$ and hence

$$P^*(I - X^*X)P \geqslant 0.$$

By Theorem 12.16, this means

$$I - \Psi_T(X)^*\Psi_T(X) \geqslant 0,$$

and so $\Psi_T(X)$ is a contraction.   □

We shall be making use of a special kind of Möbius transformation, obtained when the unitary operator T has the following form.

12.19 Definition For any contraction $B \in \mathcal{L}(H, K)$ the *Julia operator* $J(B)$ corresponding to B is the operator matrix

$$\begin{bmatrix} (I - BB^*)^{1/2} & B \\ -B^* & (I - B^*B)^{1/2} \end{bmatrix} \in \mathcal{L}(K \oplus H).$$ □

12.20 Theorem For any contraction $B \in \mathcal{L}(H, K)$, $J(B)$ is a unitary operator.

Proof.

$$J(B)^*J(B) = \begin{bmatrix} (I - BB^*)^{1/2} & -B \\ B^* & (I - B^*B)^{1/2} \end{bmatrix}\begin{bmatrix} (I - BB^*)^{1/2} & B \\ -B^* & (I - B^*B)^{1/2} \end{bmatrix}$$

$$= \begin{bmatrix} I & (I - BB^*)^{1/2}B - B(I - B^*B)^{1/2} \\ B^*(I - BB^*)^{1/2} - (I - B^*B)^{1/2}B^* & I \end{bmatrix}$$

We must prove that the two off-diagonal operators are zero. Now, for any $n \in \mathbb{N}$,

$$(BB^*)^nB = BB^*B \ldots B^*B$$

$$= B(B^*B)^n.$$

Taking a linear combination of such relations for various n we obtain

$$p(BB^*)B = Bp(B^*B)$$

for any scalar polynomial p. Since $I - BB^*$ and $I - B^*B$ are positive contractions, by Theorem 12.4 there exists a sequence (p_n) of scalar polynomials such that

$$p_n(I - BB^*)x \to (I - BB^*)^{1/2}x,$$

$$p_n(I - B^*B)x \to (I - B^*B)^{1/2}x$$

as $n \to \infty$ for all $x \in H$. Thus, since $p_n(I - BB^*)$ is a polynomial in BB^*, we have

$$(I - BB^*)^{1/2}Bx = \lim_n p_n(I - BB^*)Bx$$

$$= \lim_n Bp_n(I - B^*B)x$$

$$= B(I - B^*B)^{1/2}x.$$

That is,

$$(I - BB^*)^{1/2}B = B(I - B^*B)^{1/2}.$$

This shows that the upper off-diagonal entry in $J(B)^*J(B)$ is zero. The lower one is the adjoint of the upper, so is also zero. Hence

$$J(B)^*J(B) = I_{K \oplus H}.$$

Similar reasoning shows that

$$J(B)J(B)^* = I_{K \oplus H}. \qquad \square$$

Putting together 12.18 and 12.20 we get:

12.21 Corollary   If X and B are contractions from H to K and $I - B^*X$ is invertible in $\mathscr{L}(H)$ then

$$\Psi_{J(B)}(X) = B - (I - BB^*)^{1/2}X(I - B^*X)^{-1}(I - B^*B)^{1/2}$$

is a contraction. $\qquad \square$

12.3   Completing matrix contractions

Suppose we are given three of the four entries of a 2×2 operator matrix. We shall need to know under what circumstances we can construct the remaining entry so that the resulting operator matrix is a contraction. We can easily find some necessary conditions. Consider the operator matrix

$$T = \begin{bmatrix} P & Q \\ R & S \end{bmatrix} : H_1 \oplus H_2 \to K_1 \oplus K_2.$$

Note that

$$\begin{bmatrix} Q \\ S \end{bmatrix} : H_2 \to K_1 \oplus K_2$$

is a restriction of T. Thus, if T is a contraction, so is $\begin{bmatrix} Q \\ S \end{bmatrix}$. Similar reasoning applied to T^* shows that

$$[R \quad S] : H_1 \oplus H_2 \to K_2$$

is also a contraction. Thus

$$\begin{bmatrix} P & Q \\ R & S \end{bmatrix} \text{ a contraction} \Rightarrow \begin{bmatrix} Q \\ S \end{bmatrix}, [R \quad S] \text{ contractions}.$$

Are these conditions also sufficient? That is, if we are given Q, R and S so that $\begin{bmatrix} Q \\ S \end{bmatrix}$ and $[R \quad S]$ are contractions, can we choose P so that $\begin{bmatrix} P & Q \\ R & S \end{bmatrix}$ is a contraction?

Let us guess that the answer is yes and try to find a P that gives a contraction. Before going on to anything more elaborate we must try whether the obvious choice $P = 0$ works. In fact it does not. Consider the example

$$T = \frac{1}{\sqrt{2}}\begin{bmatrix} P & 1 \\ 1 & 1 \end{bmatrix}.$$

Here

$$\frac{1}{\sqrt{2}}\begin{bmatrix} 1 \\ 1 \end{bmatrix}, \frac{1}{\sqrt{2}}[1 \quad 1] \text{ are contractions } \mathbb{C} \to \mathbb{C}^2, \mathbb{C}^2 \to \mathbb{C},$$

but

$$T = \frac{1}{\sqrt{2}}\begin{bmatrix} 0 & 1 \\ 1 & 1 \end{bmatrix}$$

is not a contraction (apply T to the unit vector $(1/\sqrt{2})[1 \quad 1]^T$). However, if we choose $P = -1$ in this example we obtain a matrix

$$T = \frac{1}{\sqrt{2}}\begin{bmatrix} -1 & 1 \\ 1 & 1 \end{bmatrix}$$

which is actually *unitary*, and so is certainly a contraction. In general, the answer is indeed yes, but finding P is quite complicated: it makes use of all the calculations in this chapter so far.

12.22 Parrott's theorem   Let H_i, K_i be Hilbert spaces and let

$$\begin{bmatrix} Q \\ S \end{bmatrix} : H_2 \to K_1 \oplus K_2, \qquad [R \quad S] : H_1 \oplus H_2 \to K_2,$$

be contractions. There exists $P \in \mathcal{L}(H_1, K_1)$ such that

$$\begin{bmatrix} P & Q \\ R & S \end{bmatrix} : H_1 \oplus H_2 \to K_1 \oplus K_2 \tag{12.1}$$

is a contraction.

Proof. By Theorem 12.13 and Corollary 12.14, there exist contractions

$$Y : H_1 \to K_2, \qquad Z : H_2 \to K_1$$

such that

$$R = (I - SS^*)^{1/2}Y, \qquad Q = Z(I - S^*S)^{1/2}.$$

Let

$$X = \begin{bmatrix} 0 & Z \\ Y & 0 \end{bmatrix}, B = \begin{bmatrix} 0 & 0 \\ 0 & -S \end{bmatrix} : H_1 \oplus H_2 \to K_1 \oplus K_2.$$

Then X and B are contractions, and

$$I - B^*X = I - \begin{bmatrix} 0 & 0 \\ 0 & -S^* \end{bmatrix}\begin{bmatrix} 0 & Z \\ Y & 0 \end{bmatrix}$$

$$= \begin{bmatrix} I & 0 \\ S^*Y & I \end{bmatrix}.$$

Thus

$$(I - B^*X)^{-1} \text{ exists and equals } \begin{bmatrix} I & 0 \\ -S^*Y & I \end{bmatrix}.$$

By Corollary 12.21, $\Psi_{J(B)}(X)$ is a contraction. Now

$$\Psi_{J(B)}(X) = B - (I - BB^*)^{1/2}X(I - B^*X)^{-1}(I - B^*B)^{1/2}$$

$$= B - \begin{bmatrix} I & 0 \\ 0 & (I-SS^*)^{1/2} \end{bmatrix} \begin{bmatrix} 0 & Z \\ Y & 0 \end{bmatrix} \begin{bmatrix} I & 0 \\ -S^*Y & I \end{bmatrix} \begin{bmatrix} I & 0 \\ 0 & (I-S^*S)^{1/2} \end{bmatrix}$$

$$= B - \begin{bmatrix} -ZS^*Y & Z(I-S^*S)^{1/2} \\ (I-SS^*)^{1/2}Y & 0 \end{bmatrix}$$

$$= - \begin{bmatrix} -ZS^*Y & Q \\ R & S \end{bmatrix}.$$

Thus we obtain a contraction by taking $P = -ZS^*Y$.   □

12.23 Corollary Let A_{12}, A_{21} and A_{22} be given operators with $A_{ij} \in \mathscr{L}(H_j, K_i)$.

$$\inf_{P \in \mathscr{L}(H_1, K_1)} \left\| \begin{bmatrix} P & A_{12} \\ A_{21} & A_{22} \end{bmatrix} \right\| = \max \left\{ \left\| \begin{bmatrix} A_{12} \\ A_{22} \end{bmatrix} \right\|, \| [A_{21} \quad A_{22}] \| \right\}.$$

(12.2)

Proof. Since $\begin{bmatrix} A_{12} \\ A_{22} \end{bmatrix}$ is a restriction of $\begin{bmatrix} P & A_{12} \\ A_{21} & A_{22} \end{bmatrix}$, we have, for any $P \in \mathscr{L}(H_1, K_1)$,

$$\left\| \begin{bmatrix} A_{12} \\ A_{22} \end{bmatrix} \right\| \leqslant \left\| \begin{bmatrix} P & A_{12} \\ A_{21} & A_{22} \end{bmatrix} \right\|.$$

Hence, taking the infimum of the right hand side, we have

$$\left\| \begin{bmatrix} A_{12} \\ A_{22} \end{bmatrix} \right\| \leqslant \inf_P \left\| \begin{bmatrix} P & A_{12} \\ A_{21} & A_{22} \end{bmatrix} \right\|.$$

By a similar argument applied to adjoints,

$$\| [A_{21} \quad A_{22}] \| \leqslant \inf_P \left\| \begin{bmatrix} P & A_{12} \\ A_{21} & A_{22} \end{bmatrix} \right\|.$$

Combining the last two inequalities, we can assert that the left hand side of (12.2) is greater than or equal to the right hand side.

To prove the reverse inequality, let the right hand side of (12.2) be m. We can suppose $m > 0$. Then $m^{-1} \begin{bmatrix} A_{12} \\ A_{22} \end{bmatrix}, m^{-1}[A_{21} \quad A_{22}]$ are contractions and hence, by Parrott's theorem, there exists $P \in \mathscr{L}(H_1, K_1)$ such that

$$m^{-1}\begin{bmatrix} P & A_{12} \\ A_{21} & A_{22} \end{bmatrix} \text{ is a contraction. Hence}$$

$$\inf_{P}\left\|\begin{bmatrix} P & A_{12} \\ A_{21} & A_{22} \end{bmatrix}\right\| \leqslant m. \qquad \square$$

Parrott's theorem assures us that there is a single operator P which produces a 2×2 contraction: typically there will be many different possible choices of P which work, and in some applications it is useful to have a description of the set of such P. It turns out that the operators P which make (12.1) a contraction are precisely those of the form

$$P = (I - ZZ^*)^{1/2}V(I - Y^*Y)^{1/2} - ZS^*Y$$

where $V: H_1 \to K_1$ is a contraction and Y, Z have the same meanings as in the proof of Parrott's theorem. A proof of this (in the case that $\|S\| < 1$) is outlined in Problems 12.16–12.21.

The notions in the first part of this chapter, including the existence of square roots, are as old as operator theory itself. Surprisingly, the widespread use of Möbius transformations of operators is relatively recent. Parrott's theorem dates from 1978. We shall use it to prove a result of practical importance on the approximation of analytic functions, though this theorem was first proved earlier (1957) and by different methods.

12.4   Problems

12.1.   Show that the integral operator K on $L^2(0, 1)$ with kernel function $k(t, s) = \min(t, s)$ is positive (find the eigenvalues of K and invoke the spectral theorem).

12.2.   Let K be the integral operator on $L^2(-\pi, \pi)$ with kernel function k. For which of the following choice of k is K positive?

(a) $k(s, t) = \cos(s - t)$

(b) $k(s, t) = i \sin(s - t)$

(c) $k(s, t) = s + t$.

In the problems which follow H denotes an arbitrary Hilbert space.

12.3.   Let $A \in \mathcal{L}(H)$. Prove that, for all $x, y \in H$,

$$(Ax, y) = \frac{1}{4} \sum_{n=0}^{3} i^n(A(x + i^n y), x + i^n y).$$

Deduce that, if (Ax, x) is real for all $x \in H$, then A is Hermitian.

12.4.   What are the square roots of the positive operators described in Examples 12.2(ii) and (iii)?

12.5.   Show that if $A \in \mathcal{L}(H)$ and $A \geqslant 0$ then $A^n \geqslant 0$ for any $n \in \mathbb{N}$ *without* assuming the existence of a square root of A (consider even and odd n separately).

12.6. Let $A \in \mathcal{L}(H)$, let $A \geqslant 0$ and let $(Ax, x) = 0$ for some $x \in H$. Show that $Ax = 0$.

12.7. Let $A, B \in \mathcal{L}(H)$, let $A \geqslant 0$ and let $AB = BA$. Prove

(a) $p(A)B = Bp(A)$ for every scalar polynomial p;

(b) $A^{1/2}B = BA^{1/2}$;

(c) if also $B \geqslant 0$ then $AB \geqslant 0$.

12.8. Let X be a contraction on H, let $|\alpha| < 1$ and let $Y = (X - \alpha I)(I - \bar{\alpha}X)^{-1}$. Prove that Y is well defined and is a contraction.

12.9. Let $A, W \in \mathcal{L}(H)$, let $W \geqslant 0$, let $\text{Ker } W = \{0\}$ and let

$$W - AWA^* \geqslant 0.$$

Prove that $(W^{1/2})^{-1}AW^{1/2}$ is well defined on H and is a contraction.

12.10. Let $A, B, C \in \mathcal{L}(H)$ and let A be invertible. Show that

$$\begin{bmatrix} A & B \\ B^* & C \end{bmatrix} \geqslant 0 \quad \text{in } \mathcal{L}(H \oplus H)$$

if and only if $A \geqslant 0$ and $C - B^*A^{-1}B \geqslant 0$.

12.11. By diagonalization or otherwise find the square root of the 2×2 matrix

$$\begin{bmatrix} a & 1 \\ 1 & a^{-1} \end{bmatrix},$$

where $a > 0$. Hence guess and verify the square root of

$$\begin{bmatrix} A & I \\ I & A^{-1} \end{bmatrix} \in \mathcal{L}(H \oplus H)$$

where A is a positive invertible operator in $\mathcal{L}(H)$.

12.12. Show that if D is invertible and $[C \quad D]$ is a contraction then

$$\|Dx\| \geqslant \|D^{-1}\|^{-1}\|x\|, \quad \text{all } x,$$

and

$$\|C\|^2 \leqslant 1 - \|D^{-1}\|^{-2} < 1.$$

12.13. Let H, K be Hilbert spaces, let

$$T = \begin{bmatrix} A & B \\ C & D \end{bmatrix} \in \mathcal{L}(K \oplus H)$$

be unitary and let D be invertible.

(a) Show that the Möbius transformation Ψ_T has domain containing the set $\mathscr{C}$ of contractions from H to K.

(b) Let

$$R = \begin{bmatrix} A^* & -C^* \\ -B^* & D^* \end{bmatrix}.$$

Prove that R is unitary and that Ψ_R is the inverse mapping of $\Psi_T : \mathscr{C} \to \mathscr{C}$. Deduce that, for any strict contraction $B \in \mathscr{L}(H, K)$, $\Psi_{J(B)}$ is an involution on $\mathscr{C}$ (i.e. $\Psi_{J(B)}^{-1} = \Psi_{J(B)}$).

12.14. Find a unitary operator U such that $\Psi_T(X) = \Psi_U(X^*)^*$, and hence deduce Corollary 12.17 from Theorem 12.16.

12.15. Find

$$\inf_{a \in C} \left\| \begin{bmatrix} a & 1 & 2 \\ 1 & 2 & 0 \\ 2 & 0 & 0 \end{bmatrix} \right\|$$

and a value of a for which the infimum is attained.

12.16. Prove that, if B is a strict contraction,

$$\Psi_{J(B)}(X) = (I - BB^*)^{-1/2}(B - X)(I - B^*X)^{-1}(I - B^*B)^{1/2}$$
$$= (I - BB^*)^{1/2}(I - XB^*)^{-1}(B - X)(I - B^*B)^{-1/2}.$$

12.17. Let

$$B = \begin{bmatrix} 0 & Z \\ Y & 0 \end{bmatrix} : H_1 \oplus H_2 \to K_1 \oplus K_2$$

be a contraction. Show that, for any $V \in \mathscr{L}(H_1, K_1)$,

$$T = \begin{bmatrix} -V & 0 \\ 0 & 0 \end{bmatrix}$$

is in the domain of $\Psi_{J(B)}$, and

$$\Psi_{J(B)}(T) = \begin{bmatrix} (I - ZZ^*)^{1/2}V(I - Y^*Y)^{1/2} & Z \\ Y & 0 \end{bmatrix}.$$

12.18. Let

$$X = \begin{bmatrix} W & Z \\ Y & 0 \end{bmatrix}, B = \begin{bmatrix} 0 & 0 \\ 0 & -S \end{bmatrix} : H_1 \oplus H_2 \to K_1 \oplus K_2$$

and let B be a contraction. Prove that $X \in$ domain $\Psi_{J(B)}$ and that

$$\Psi_{J(B)}(X) = - \begin{bmatrix} W - ZS^*Y & Z(I - S^*S)^{1/2} \\ (I - SS^*)^{1/2}Y & S \end{bmatrix}.$$

12.19. Let Q, R, S, Y and Z be as in the statement and proof of Parrott's theorem. Deduce from Problems 12.17 and 12.18 that

$$\begin{bmatrix} P & Q \\ R & S \end{bmatrix} \text{ is a contraction}$$

whenever

$$P = (I - ZZ^*)^{1/2}V(I - Y^*Y)^{1/2} - ZS^*Y$$

for some contraction $V \in \mathscr{L}(H_1, K_1)$.

12.20. Let

$$\begin{bmatrix} W & Z \\ Y & 0 \end{bmatrix} : H_1 \oplus H_2 \to K_1 \oplus K_2$$

be a contraction. Prove that there exists a contraction $V_1 : H_1 \to K_1 \oplus K_2$ such that

$$\begin{bmatrix} W \\ Y \end{bmatrix} = \begin{bmatrix} (I - ZZ^*)^{1/2} & 0 \\ 0 & I \end{bmatrix} V_1 .$$

Deduce that there exists a contraction $V : H_1 \to K_1$ such that

$$W = (I - ZZ^*)^{1/2} V (I - Y^*Y)^{1/2}.$$

12.21. (Converse of Problem 12.19.) Let Q, R, S, Y and Z be as in Problem 12.19, and suppose that S is a strict contraction. Using Problems 12.16 and 12.18 show that if P is such that

$$T = \begin{bmatrix} P & Q \\ R & S \end{bmatrix}$$

is a contraction, then

$$\Psi_{J(B)}(-T) = \begin{bmatrix} P + ZS^*Y & Z \\ Y & 0 \end{bmatrix}.$$

Deduce that there exists a contraction $V : H_1 \to K_1$ such that

$$P = (I - ZZ^*)^{1/2} V (I - Y^*Y)^{1/2} - ZS^*Y.$$

(This proof can be modified so as to work when $\|S\| = 1$: one uses $\Psi_{J(rB)}$, $0 < r < 1$, and then lets $r \to 1$. It requires basic facts about manipulating functions of operators – an important topic, but one we do not cover here.)

13

Hardy spaces

The Fourier series $\sum c_n e^{in\theta}$ can be converted into something which looks like a Laurent expansion, $\sum c_n z^n$, by the change of variable $z = e^{i\theta}$. In this chapter we discuss the interplay between complex analysis and the Fourier analysis of periodic functions on $\mathbb{R}$. A 2π-periodic function f on $\mathbb{R}$ can be identified in an obvious way with a function F on the unit circle $\partial\mathbb{D}$: we simply let $F(e^{i\theta}) = f(\theta)$. We shall therefore think in terms of functions on $\partial\mathbb{D}$. Certain of these functions extend naturally to analytic functions in $\mathbb{D}$, and we shall be concerned with the properties of their analytic extensions and the relationship between the original function and the extended one.

First we introduce the relevant spaces of functions. We shall be integrating with respect to arc length on the unit circle, normalized so that the length of the circle is 1. This corresponds to integrating with respect to $d\theta/2\pi$ on $(-\pi, \pi)$ under the correspondence $\theta \leftrightarrow e^{i\theta}$.

13.1 Definitions L^2 denotes the Hilbert space of square-integrable Lebesgue measurable $\mathbb{C}$-valued functions on the unit circle $\partial\mathbb{D}$, with pointwise algebraic operations and inner product

$$(f, g) = \frac{1}{2\pi} \int_{-\pi}^{\pi} f(e^{i\theta}) \overline{g(e^{i\theta})} \, d\theta \tag{13.1}$$

$$= \frac{1}{2\pi i} \int_{\partial\mathbb{D}} f(z) \overline{g(z)} \, \frac{dz}{z}. \tag{13.2}$$

L^∞ denotes the Banach space of essentially bounded Lebesgue-measurable $\mathbb{C}$-valued functions on $\partial\mathbb{D}$ with pointwise algebraic operations and essential supremum norm:

$$\|f\|_\infty = \operatorname*{ess\,sup}_{|z|=1} |f(z)|. \qquad \square \tag{13.3}$$

To say that f is square-integrable on $\partial\mathbb{D}$ means

$$\int_{-\pi}^{\pi} |f(e^{i\theta})|^2\, d\theta < \infty.$$

L^2 is essentially the same space as $L^2(-\pi, \pi)$, as described in Example 3.5, and the discussion there applies here also. Thus, functions which differ only on a set of measure zero are to be regarded as the same element of L^2. The space RL^2 of Example 1.16 is a subspace of L^2, with the same inner product, and those virtuous readers who have done all the exercises relating to RL^2 will already have some familiarity with L^2. The equality of the two expressions (13.1) and (13.2) for the inner product follows from the definition of a contour integral, with the parametrization $z(\theta) = e^{i\theta}$, $-\pi < \theta \leqslant \pi$ of the unit circle. We shall often omit the '$\partial\mathbb{D}$' beneath the integral sign in this context.

A function is said to be *essentially bounded* if it is bounded on the complement of a set of measure zero. For present purposes there is little harm in thinking of L^∞ as the space of *all* bounded complex-valued functions on the unit circle. In the definition of the norm the word 'essential' is there because, once again, functions agreeing except on a set of measure zero are identified. Thus, if we redefine a function at a single point, giving it a very large value, we do not want it to change the norm of the function. This leads us to say that a number M is an *essential upper bound* for a function $g: \partial\mathbb{D} \to \mathbb{R}$ if

$$\{z \in \partial\mathbb{D} : g(z) > M\}$$

is a set of measure zero, in the sense explained in Example 3.5. Then we define

$$\operatorname*{ess\,sup}_{|z|=1} |f(z)|$$

$$= \inf\{M : M \text{ is an essential upper bound for } |f(z)| \text{ on } \partial\mathbb{D}\}.$$

For continuous functions the essential supremum is the same as the supremum, and so this technicality can usually be ignored. We have to accept from measure theory that the normed space L^∞ is complete.

13.2 Exercise Prove that L^∞ is a proper subspace of L^2 and that

$$\|f\|_{L^2} \leqslant \|f\|_\infty$$

for all $f \in L^\infty$.   □

In view of the correspondence between L^2 and $L^2(-\pi, \pi)$ we can restate Theorem 5.1 on the completeness of the sequence $(e^{in\theta})$.

13.3 Theorem The sequence $(z^n)_{n=-\infty}^{\infty}$ is a complete orthonormal sequence in L^2.   □

Strictly speaking, we should say something like 'the sequence $(f_n)_{n=-\infty}^{\infty}$, where $f_n(z) = z^n$: however, this slight abuse of notation is convenient and will not normally confuse.

We shall write $\hat{f}(n)$ for the nth Fourier coefficient of f with respect to (z^n). Thus, for $f \in L^2$,

$$\hat{f}(n) = (f, z^n)$$

$$= \frac{1}{2\pi} \int_{-\pi}^{\pi} f(e^{i\theta}) e^{-in\theta} d\theta.$$

By Theorems 4.14 and 13.3, if $f \in L^2$ and $\hat{f}(n) = a_n$, $n \in \mathbb{Z}$, then

$$f(z) = \sum_{n=-\infty}^{\infty} a_n z^n$$

in the sense of L^2 convergence. Which functions f have obvious extensions to analytic functions in $\mathbb{D}$? A promising try is to take the functions for which $a_{-1}, a_{-2}, \dots$ are all zero, for then

$$f(z) = \sum_{n=0}^{\infty} a_n z^n$$

on $\partial \mathbb{D}$, and the same formula defines an analytic function in $\mathbb{D}$. Let us make this idea precise.

13.4 Definition   Let $p = 2$ or ∞. H^p is the closed subspace

$$\{f \in L^p : \hat{f}(n) = 0 \text{ for } n < 0\}$$

of L^p, with the restriction of the norm of L^p.   □

H^2 and H^∞ are called *Hardy spaces*. H^2 is a Hilbert space and H^∞ a Banach space. Clearly $H^\infty \subseteq H^2$. RH^2 (see Examples 1.16) comprises the rational functions of z which belong to H^2, regarded as functions on $\partial \mathbb{D}$.

13.5 Theorem   Let $f \in H^2$ and let $\hat{f}(n) = a_n$, $n \in \mathbb{Z}$. Then the function

$$\tilde{f}(z) = \sum_{n=0}^{\infty} a_n z^n \tag{13.4}$$

is defined and analytic in $\mathbb{D}$.

Proof. Since $f \in L^2$, $\sum_0^\infty |a_n|^2 < \infty$ by Bessel's inequality. Hence $|a_n| \to 0$ as $n \to \infty$, and so

$$\limsup_{n \to \infty} |a_n|^{1/n} \leqslant 1$$

(indeed, $|a_n|^{1/n} < 1$ for all but finitely many n). Hence the radius of convergence of the power series (13.4) is at least 1. Thus $\tilde{f}$ is well defined on $\mathbb{D}$. Furthermore, in any disc $\{z : |z| \leqslant r\}$, where $0 < r < 1$, the series (13.4) converges uniformly, by Weierstrass' M-test. Thus $\tilde{f}$ is a uniform limit of polynomials on $\{z: |z| < r\}$, and so is analytic there. Since this holds for all $r < 1$, $\tilde{f}$ is analytic on $\mathbb{D}$.   □

Thus, for $f \in H^2$ with $\hat{f}(n) = a_n$ we have

$$f(z) = \sum_0^\infty a_n z^n \qquad (13.5)$$

on $\partial \mathbb{D}$ in the sense of L^2 convergence, and

$$\tilde{f}(z) = \sum_0^\infty a_n z^n$$

uniformly on compact subsets of $\mathbb{D}$. Although f and $\tilde{f}$ are given by the same formula, they are distinct objects: their domains of definition are disjoint. How, then, can we express the relationship between f and $\tilde{f}$ without invoking Fourier coefficients? Think first of the case of rational f. Here the power series $\sum a_n z^n$ will converge to an analytic function in the open disc of centre 0 and radius

$$\min\{|\lambda| : \lambda \text{ is a pole of } f\},$$

a disc which properly contains the closure of $\mathbb{D}$. It follows that the function on clos $\mathbb{D}$ which agrees with f on $\partial \mathbb{D}$ and $\tilde{f}$ on $\mathbb{D}$ is continuous. We therefore have, for $f \in RH^2$ and $z \in \partial \mathbb{D}$,

$$f(z) = \lim_{r \to 1-} \tilde{f}(rz). \qquad (13.6)$$

Thus, if we know $\tilde{f}$ we can recover f by taking 'radial limits'. On the other hand, if we know f we can calculate $\tilde{f}(a)$, for $a \in \mathbb{D}$, by means of Cauchy's integral formula. We can therefore get from either f or $\tilde{f}$ to the other in a direct manner.

This argument does not work for a general, non-rational, function in H^2. The power series (13.5) need not converge at every point of $\partial \mathbb{D}$, and indeed, the limit on the right hand side of (13.6) need not exist for every $z \in \partial \mathbb{D}$. It is useful to know an example of a function for which these things go wrong: in trying to decide whether a particular property of rational functions holds for general H^2 functions it is very helpful to try it out on the following example.

13.6 Example   A singular inner function. Let

$$\varphi(z) = \frac{z+1}{z-1}.$$

Then

$$\mathrm{Re}\,\varphi(z) = \frac{1}{2}\left\{\frac{z+1}{z-1} + \frac{\bar{z}+1}{\bar{z}-1}\right\}$$

$$= -\frac{1 - |z|^2}{|z-1|^2}.$$

It follows that φ maps $\mathbb{D}$ to the open left half plane and $\partial\mathbb{D}$ to the imaginary axis. Now the exponential function maps the left half plane to $\mathbb{D}$ and the imaginary axis to $\partial\mathbb{D}$. Hence the function

$$F(z) = \exp\frac{z+1}{z-1}$$

maps $\mathbb{D}$ to $\mathbb{D}$ and $\partial\mathbb{D}\backslash\{1\}$ to $\partial\mathbb{D}$. Let f be the restriction of F to $\partial\mathbb{D}\backslash\{1\}$: f is well defined, bounded and continuous on the complement of a set of measure zero in $\partial\mathbb{D}$, hence is a member of L^{∞}. F is analytic in $\mathbb{D}$ and so has a Taylor expansion

$$F(z) = \sum_0^{\infty} a_n z^n$$

valid there. The expansion is not valid on the unit circle itself, since F has an essential singularity at 1, but it is nevertheless true that the Fourier series of $f(e^{i\theta})$ is

$$f(e^{i\theta}) \sim \sum_0^{\infty} a_n e^{in\theta}.$$

We shall prove this shortly (Theorem 13.11). It follows that $f \in H^{\infty}$ and $\tilde{f}$ agrees with F on $\mathbb{D}$. Since F is continuous on $\mathbb{C}\backslash\{1\}$, the relation

$$f(z) = \lim_{r\to 1-} \tilde{f}(rz)$$

holds for $z \in \partial\mathbb{D}\backslash\{1\}$, but when $z = 1$ the limit on the right hand side is zero while the left hand side is undefined.

A function $g \in H^{\infty}$ is said to be *inner* if $|g(z)| = 1$ almost everywhere on $\partial\mathbb{D}$, and to be *singular inner* if it is inner and is not divisible by any rational inner function in H^{∞}. The present f is the simplest example of a singular inner function. $\square$

13.1   Poisson's kernel

Despite Example 13.6, we shall soon see that f *can* be recovered from $\tilde{f}$ by the process of taking radial limits. First, though, we need an expression for $\tilde{f}$ in terms of f.

13.7   *Theorem*   Let $f \in H^2$. For $0 \leqslant r < 1$ and $\theta \in \mathbb{R}$,

$$\tilde{f}(re^{i\theta}) = \frac{1}{2\pi} \int_{-\pi}^{\pi} f(e^{it}) P_r(\theta - t) \, dt \tag{13.7}$$

where

$$P_r(\theta) = \frac{1 - r^2}{1 - 2r\cos\theta + r^2}. \tag{13.8}$$

Proof. Let $\hat{f}(n) = a_n$ for $n \in \mathbb{Z}$. Since $f \in H^2$, $a_n = 0$ when $n < 0$, and so, by definition of $\hat{f}$,

$$\tilde{f}(re^{i\theta}) = \sum_0^\infty a_n r^n e^{in\theta}$$

$$= \sum_{-\infty}^\infty a_n r^{|n|} e^{in\theta}$$

$$= \sum_{-\infty}^\infty \frac{1}{2\pi} \int_{-\pi}^\pi f(t) e^{-int} \, dt \; r^{|n|} e^{in\theta}$$

$$= \frac{1}{2\pi} \sum_{-\infty}^\infty \int_{-\pi}^\pi f(t) r^{|n|} e^{in(\theta - t)} \, dt$$

$$= \frac{1}{2\pi} \int_{-\pi}^\pi f(t) \sum_{-\infty}^\infty r^{|n|} e^{in(\theta - t)} \, dt. \tag{13.9}$$

The interchange of summation and integration here is allowable since, for fixed $r < 1$, the infinite series in the last line is a uniformly convergent series of functions on $(-\pi, \pi)$, by Weierstrass' M-test. We have, putting $z = re^{i\theta}$,

$$\sum_{-\infty}^\infty r^{|n|} e^{in\theta} = \sum_1^\infty r^n e^{-in\theta} + \sum_0^\infty r^n e^{in\theta}$$

$$= \sum_1^\infty \bar{z}^n + \sum_0^\infty z^n$$

$$= \frac{\bar{z}}{1 - \bar{z}} + \frac{1}{1 - z}$$

$$= \frac{1 - z\bar{z}}{(1 - z)(1 - \bar{z})} \tag{13.10}$$

$$= \frac{1 - |z|^2}{1 - (z + \bar{z}) + |z|^2}$$

$$= \frac{1 - r^2}{1 - 2r \cos \theta + r^2}$$

$$= P_r(\theta).$$

Thus (13.9) can be written

$$\tilde{f}(re^{i\theta}) = \frac{1}{2\pi} \int_{-\pi}^\pi f(t) P_r(\theta - t) \, dt. \qquad \square$$

The function P_r is called *Poisson's kernel*, and the right hand side of (13.7) is called the *Poisson integral* of f.

13.8 Lemma $P_r(\theta) > 0$ for all $r \in [0, 1)$, $\theta \in \mathbb{R}$, and

$$\frac{1}{2\pi} \int_{-\pi}^\pi P_r(\theta) \, d\theta = 1.$$

Proof. Going back to (13.10) we find

$$P_r(\theta) = \frac{1 - |z|^2}{|1 - z|^2} > 0$$

where $z = re^{i\theta}$. Putting $w = e^{i\theta}$ we have

$$P_r(\theta) = \frac{1 - r^2}{(1 - rw)(1 - r\bar{w})}$$

$$= \frac{w(1 - r^2)}{(1 - rw)(w - r)}.$$

Hence

$$\frac{1}{2\pi} \int_{-\pi}^{\pi} P_r(\theta) \, d\theta = \frac{1}{2\pi i} \int_{\partial D} \frac{w(1 - r^2)}{(1 - rw)(w - r)} \frac{dw}{w}$$

$$= \frac{1}{2\pi i} \int_{\partial D} \frac{1 - r^2}{(1 - rw)(w - r)} \, dw.$$

The integrand is analytic on the closed unit disc except for a simple pole with residue 1 at $w = r$. Hence, by the residue theorem, the right hand side equals 1. $\qquad\square$

Our next enterprise is to show that H^∞, which we have defined as a space of functions on the circle, can be identified with the space of bounded analytic functions in $\mathbb{D}$. We shall do this in stages.

13.9 Theorem   For any $f \in H^\infty$ and $z \in \mathbb{D}$,

$$|\tilde{f}(z)| \leqslant \|f\|_\infty,$$

where $\tilde{f}$ is defined by (13.4). Hence $\tilde{f}$ is bounded and analytic in $\mathbb{D}$.
Proof. Let $z = re^{i\theta}$, $r, \theta \in \mathbb{R}$. By Theorem 13.7,

$$\tilde{f}(re^{i\theta}) = \frac{1}{2\pi} \int_{-\pi}^{\pi} f(t) P_r(\theta - t) \, dt.$$

By the positivity of P_r,

$$|\tilde{f}(re^{i\theta})| \leqslant \frac{1}{2\pi} \int_{-\pi}^{\pi} |f(t)| P_r(\theta - t) \, dt$$

$$\leqslant \|f\|_\infty \cdot \frac{1}{2\pi} \int_{-\pi}^{\pi} P_r(\theta - t) \, dt. \tag{13.11}$$

Making the change of variable $u = \theta - t$, θ being fixed, we have

$$\frac{1}{2\pi} \int_{-\pi}^{\pi} P_r(\theta - t) \, dt = \frac{1}{2\pi} \int_{-\pi + \theta}^{\pi + \theta} P_r(u) \, du$$

$$= 1. \tag{13.12}$$

The last line follows from Lemma 13.8 and the fact that P_r is 2π-periodic (obvious from (13.8)), so that integrating P_r over any interval of length 2π will give the same result. Combining (13.11) and (13.12) gives

$$|\tilde{f}(z)| \leqslant \|f\|_\infty.$$ □

13.2   Fatou's theorem

So an element f of H^∞ gives rise to a bounded analytic function $\tilde{f}$ on $\mathbb{D}$. Can we go the other way? If we have a function F, bounded and analytic in $\mathbb{D}$, can we produce a function $f \in H^\infty$ such that $\tilde{f} = F$? We might try defining f by

$$f(z) = \lim_{r \to 1-} F(rz) \quad \text{for } z \in \partial\mathbb{D},$$

but there are discouraging examples which show that the limit on the right hand side does not necessarily exist. In Example 13.6 the radial limit relation only fails to hold at the single point 1 of $\partial\mathbb{D}$, and since H^∞ functions only need to be defined almost everywhere it remains possible that the limit will exist often enough to give us the desired function f. This may look optimistic, but it turns out to be true.

13.10 Fatou's theorem   Let $f \in H^2$. For almost all $z \in \partial\mathbb{D}$,

$$\lim_{r \to 1-} \tilde{f}(rz) = f(z).$$ (13.13)

Proof. If f is constant and equal to c on $\partial\mathbb{D}$ then it is immediate from the definition of $\tilde{f}$ that $\tilde{f}$ is identically equal to c on $\mathbb{D}$, and hence (13.13) is valid for all $z \in \mathbb{D}$. We may therefore subtract a constant from both sides of (13.13) to reduce to the case that

$$\int_{-\pi}^{\pi} f(e^{i\theta}) \, d\theta = 0.$$ (13.14)

Define a continuous function g on $\mathbb{R}$ by

$$g(t) = \int_{-\pi}^{t} f(e^{i\xi}) \, d\xi.$$

By (13.14), $g(\pi) = 0 = g(-\pi)$, and g is a 2π-periodic function. We require two facts from integration theory (see Rudin, 1966, p. 165). Although g need not be differentiable at every point of $\mathbb{R}$ (recall Problem 10.1) it *is* differentiable almost everywhere, and

$$g'(t) = f(e^{it})$$

for almost all $t \in \mathbb{R}$. The second fact is that the formula for integration by parts applies. Thus, starting from Theorem 13.7 we have

$$\tilde{f}(re^{i\theta}) = \frac{1}{2\pi} \int_{-\pi}^{\pi} P_r(\theta - t) f(e^{it}) \, dt$$

$$= \frac{1}{2\pi} [P_r(\theta - t)g(t)]_{-\pi}^{\pi} + \frac{1}{2\pi} \int_{-\pi}^{\pi} P_r'(\theta - t)g(t) \, dt.$$

The first term on the right hand side is zero, and on making the substitution $u = \theta - t$ we find

$$\tilde{f}(re^{i\theta}) = \frac{1}{2\pi} \int_{-\pi + \theta}^{\pi + \theta} P_r'(u) g(\theta - u)\, du$$

$$= \frac{1}{2\pi} \int_{-\pi}^{\pi} P_r'(u) g(\theta - u)\, du,$$

the last step holding because the integrand is 2π-periodic, so that its integral over any interval of length 2π has the same value. Differentiating (13.8) we have

$$P_r'(\theta) = -\frac{2r(1 - r^2) \sin \theta}{(1 - 2r \cos \theta + r^2)^2}.$$

Clearly $P_r'(-\theta) = -P_r'(\theta)$. Thus, putting $t = -u$ in the first of the following integrals and $t = u$ in the second we obtain

$$\tilde{f}(re^{i\theta}) = \frac{1}{2\pi} \int_{-\pi}^{0} + \int_{0}^{\pi} P_r'(u) g(\theta - u)\, du$$

$$= \frac{1}{2\pi} \int_{0}^{\pi} P_r'(t)\{g(\theta - t) - g(\theta + t)\}\, dt.$$

Let

$$K_r(t) = -\frac{2}{r} \sin t \, P_r'(t) \tag{13.15}$$

$$= \frac{4(1 - r^2) \sin^2 t}{(1 - 2r \cos t + r^2)^2}. \tag{13.16}$$

Then, for fixed θ,

$$\tilde{f}(re^{i\theta}) = \frac{r}{2\pi} \int_{0}^{\pi} K_r(t) G(t)\, dt \tag{13.17}$$

where

$$G(t) = \frac{g(\theta + t) - g(\theta - t)}{2 \sin t}.$$

If g is differentiable at θ then, by l'Hôpital's rule,

$$\lim_{t \to 0} G(t) = g'(\theta). \tag{13.18}$$

Similarly, if g is differentiable at $\theta + \pi$, we have (by virtue of the 2π-periodicity of g),

$$\lim_{t \to \pi} G(t) = \lim_{\tau \to 0} \frac{g(\theta + \pi - \tau) - g(\theta - \pi + \tau)}{2 \sin(\pi - \tau)}$$

$$= -g'(\theta + \pi). \tag{13.19}$$

Let E denote the set of $\theta \in \mathbb{R}$ such that g is differentiable at θ and at $\theta + \pi$. E

is the complement of a set of measure zero, and for $\theta \in E$, (13.18), (13.19) hold. It follows that, for $\theta \in E$ there exists $M > 0$ (depending on θ) such that

$$|G(t) - g'(\theta)| < M, \qquad 0 < t < \pi. \tag{13.20}$$

It is immediate from (13.16) that the functions $K_r, 0 \leqslant r < 1$, have the properties

(i) $K_r(t) \geqslant 0, t \in \mathbb{R}$;

(ii) $K_r(t) \to 0$ as $r \to 1$, uniformly on (δ, π) for any $\delta \in (0, \pi)$.

Furthermore, by evaluating a contour integral (see Problem 13.8) we can show that

(iii) $\displaystyle\int_0^\pi K_r(t) \, dt = 2\pi, \qquad 0 \leqslant r < 1.$

We shall use these properties of K_r much as we did those of the Fejér kernel in Theorem 5.1. From (iii) and (13.17), for $r > 0$,

$$\frac{1}{r} \tilde{f}(re^{i\theta}) - g'(\theta) = \frac{1}{2\pi} \int_0^\pi K_r(t)(G(t) - g'(\theta)) \, dt. \tag{13.21}$$

We must show that the left hand side tends to zero as $r \to 1-$.

Let $\varepsilon > 0$. Consider $\theta \in E$. By (13.18) there exists $\delta \in (0, \pi)$ such that

$$|G(t) - g'(\theta)| < \frac{\varepsilon}{2}, \qquad 0 < t < \delta. \tag{13.22}$$

Then (13.21) can be written

$$\frac{1}{r} \tilde{f}(re^{i\theta}) - g'(\theta) = \frac{1}{2\pi} \left[\int_0^\delta + \int_\delta^\pi \right] K_r(t)(G(t) - g'(\theta)) \, dt$$

$$= I_1 + I_2, \quad \text{say.}$$

Here

$$|I_1| \leqslant \frac{1}{2\pi} \int_0^\delta K_r(t) |G(t) - g'(\theta)| \, dt,$$

and so, in view of (13.22),

$$|I_1| \leqslant \frac{\varepsilon}{2} \cdot \frac{1}{2\pi} \int_0^\delta K_r(t) \, dt \leqslant \frac{\varepsilon}{2}.$$

By property (ii) of K_r, there exists $r_0 \in (0, 1)$ such that $r_0 < r < 1$ implies

$$|K_r(t)| < \frac{\varepsilon}{M}, \qquad \delta < t < \pi.$$

Hence, when $r_0 < r < 1$,

$$|I_2| \leqslant \frac{1}{2\pi} \int_\delta^\pi K_r(t) |G(t) - g'(\theta)| \, dt$$

$$\leqslant \frac{\varepsilon}{M} \cdot \frac{1}{2\pi} \int_\delta^\pi |G(t) - g'(\theta)| \, dt$$

$$\leqslant \frac{\varepsilon}{M} \cdot \frac{1}{2\pi} M \cdot \pi = \frac{\varepsilon}{2}.$$

We have shown that if $\theta \in E$ then for all $\varepsilon > 0$ there exists $r_0 \in (0, 1)$ such that $r_0 < r < 1$ implies

$$\left| \frac{1}{r} \tilde{f}(re^{i\theta}) - g'(\theta) \right| \leqslant |I_1 + I_2|$$

$$\leqslant \varepsilon.$$

That is,

$$\lim_{r \to 1-} \frac{1}{r} \tilde{f}(re^{i\theta}) = g'(\theta) = f(e^{i\theta}),$$

and so

$$\lim_{r \to 1-} \tilde{f}(re^{i\theta}) = f(e^{i\theta})$$

whenever $\theta \in E$; that is, for almost all $\theta \in \mathbb{R}$. $\square$

The full statement of Fatou's theorem is stronger than what is proved here. It applies to a wider class of functions f and is valid for a more general kind of limit than radial limits. Details are in Hoffman (1962).

13.11 Theorem If F is a bounded analytic function in $\mathbb{D}$ there exists a unique $f \in H^\infty$ such that $\tilde{f} = F$: moreover

$$\| f \|_\infty = \sup_{z \in \mathbb{D}} |F(z)|.$$

Proof. Let

$$M = \sup_{z \in \mathbb{D}} |F(z)|$$

and let the Taylor expansion of F be

$$F(z) = \sum_0^\infty a_n z^n, \tag{13.19}$$

valid in $\mathbb{D}$. Define a function f_r on $\partial \mathbb{D}$, $0 \leqslant r < 1$, by

$$f_r(z) = F(rz).$$

Then f_r is continuous and bounded by M in modulus. It follows that $f_r \in L^2$ and $\| f_r \|_{L^2} \leqslant M$. Since

$$f_r(z) = \sum_0^\infty a_n r^n z^n$$

uniformly on $\partial \mathbb{D}$, it is clear that the nth Fourier coefficient of f_r is $a_n r^n$, so that Bessel's inequality gives

$$\sum_0^k |a_n|^2 r^{2n} \leqslant M^2$$

for any $k \in \mathbb{N}$. As this holds for every $r \in [0, 1)$ we may let $r \to 1-$ to obtain

$$\sum_0^k |a_n|^2 \leqslant M^2.$$

Now let $k \to \infty$:

$$\sum_0^\infty |a_n|^2 \leqslant M^2.$$

Thus, by Theorem 4.11, the formula

$$f(z) = \sum_0^\infty a_n z^n$$

defines an element f of L^2. Since the negative Fourier coefficients of f are zero, $f \in H^2$. It is clear from (13.19) and the definition of $\tilde{f}$ that $\tilde{f} = F$.

We now have an $f \in H^2$ to which we can apply Fatou's theorem:

$$f(z) = \lim_{r \to 1-} \tilde{f}(rz)$$

$$= \lim_{r \to 1-} F(rz)$$

for all $z \in \partial \mathbb{D} \backslash E$, where E is some set of measure zero in $\partial \mathbb{D}$. I claim that M is an essential upper bound for $|f|$. Suppose not: then

$$\{z \in \partial \mathbb{D} : |f(z)| > M\}$$

is a set of positive measure, and hence cannot be a subset of E. There consequently exists $z \in \partial \mathbb{D}$ such that

$$|f(z)| > M$$

and

$$\lim_{r \to 1-} F(rz) = f(z).$$

Since $|F(rz)| \leqslant M$ for $0 \leqslant r < 1$, by choice of M, this is a contradiction. Hence

$$\|f\|_\infty \leqslant M = \sup_{z \in \mathbb{D}} |F(z)|,$$

and $f \in H^\infty$ (we may re-define f to be identically zero on the set E of measure zero to ensure that f is bounded on $\partial \mathbb{D}$). On the other hand, Theorem 13.9 tells us that

$$|F(z)| = |\tilde{f}(z)| \leqslant \|f\|_\infty$$

for every $z \in \mathbb{D}$, and so

$$\sup_{z \in \mathbb{D}} |F(z)| = \|f\|_\infty. \qquad \square$$

The correspondence $f \leftrightarrow \tilde{f}$ is thus an isometric isomorphism between H^∞, with essential supremum norm, and the Banach space of bounded analytic functions on $\mathbb{D}$ with supremum norm. Accordingly we shall henceforward ignore the distinction between these two spaces. We shall drop the notation $\tilde{f}$ and regard an element f of H^∞ either as an essentially bounded function on the circle or as a bounded analytic function in $\mathbb{D}$, whichever suits us best at the time.

There is an analogous result for H^2. Since H^2 is a larger class of functions on $\partial\mathbb{D}$ than H^∞, the set of $\tilde{f}$ with $f \in H^2$ is a set of analytic functions on $\mathbb{D}$ which contains all bounded analytic functions, but also contains some unbounded ones. One can ask which analytic functions on $\mathbb{D}$ belong to this set. The answer is that it comprises those analytic functions F such that

$$\sup_{0 \leqslant r < 1} \int_{-\pi}^{\pi} |F(re^{i\theta})|^2 \, d\theta < \infty.$$

We shall not use this fact and so will not prove it. However, it is as well to be aware that H^2, like H^∞, can be thought of as a Banach space of analytic functions on $\mathbb{D}$. A proof can be found in Hoffman (1962) or Rudin (1966).

13.12   Theorem   If $f \in H^2$ and $g \in H^\infty$ then $fg \in H^2$.

Proof. We can see either from Theorem 13.11 or directly from the definition of H^∞ that the function $z^j g(z) \in H^\infty$ for $j = 0, 1, 2, \dots$ Hence $pg \in H^\infty$ for any polynomial p. In particular $f_n g \in H^\infty$ where

$$f_n(z) = \sum_{j=0}^{n} \hat{f}(j) z^j$$

and $n \in \mathbb{N}$. Since $H^\infty \subset H^2$, $f_n g \in H^2$. Now it is clear that fg is square integrable on $\partial\mathbb{D}$, hence belongs to L^2. And

$$\| fg - f_n g \|_{L^2}^2 = \frac{1}{2\pi} \int_{-\pi}^{\pi} |f - f_n|^2 |g|^2 \, d\theta$$

$$\leqslant \| g \|_\infty^2 \| f - f_n \|_{L^2}^2.$$

Since $f_n \to f$ in L^2 it follows that $f_n g \to fg$ in L^2. Since H^2 is closed in L^2, $fg \in H^2$. $\square$

13.3   Zero sets of H^2 functions

If a rational H^2 function vanishes infinitely often on $\partial\mathbb{D}$ then its zeros have a limit point in $\partial\mathbb{D}$, and hence the function is identically zero, since the zeros of a non-constant analytic function are isolated. What of the zeros of a general H^2 function in $\partial\mathbb{D}$? We can re-define such a function arbitrarily on any set of measure zero, so the class of zero sets of H^2 functions contains all set of measure zero. We show now that it contains no other subsets of $\partial\mathbb{D}$, other than sets whose complements have measure zero. In the following, think of 'measure' as 'length' if you are unfamiliar with it.

13.13 Theorem   If $f \in H^2$ and f is zero on a set of positive measure in $\partial\mathbb{D}$ then $f = 0$.

Proof. Suppose that f vanishes on a set E of positive measure in $\partial\mathbb{D}$ but f is not the zero function (i.e. f is not equal to zero almost everywhere). Then some Fourier coefficient in the expansion

$$f(z) \sim \sum_0^\infty a_n z^n$$

is non-zero, and by dividing by a suitable power of z we can reduce to the case that $a_0 \neq 0$. Let C be the set of functions of the form

$$f(z)(1 + b_1 z + \cdots + b_n z^n)$$

with $n \in \mathbb{N}$, $b_1, \ldots, b_n \in \mathbb{C}$. C is a convex subset of H^2: so therefore is its closure clos C. By Theorem 3.8, clos C contains an element g of smallest norm. Every element of C (and hence of clos C) has its first Fourier coefficient equal to a_0, and so is not the zero function. In particular, $g \neq 0$.

For any $\lambda \in \mathbb{C}$ and $n \in \mathbb{N}$ we have

$$\|g + \lambda z^n g\|^2 = \|g\|^2 (1 + |\lambda|^2) + 2 \operatorname{Re} \frac{\lambda}{2\pi i} \int |g(z)|^2 z^n \frac{dz}{z}.$$

Since $(1 + \lambda z^n)g \in \operatorname{clos} C$, this quantity attains its minimum when $\lambda = 0$. This implies that

$$\int_{-\pi}^{\pi} |g(e^{i\theta})|^2 e^{in\theta} \, d\theta = 0$$

for all $n \in \mathbb{N}$. On taking complex conjugates we obtain the same conclusion for all negative integers n. Thus all Fourier coefficients of $|g|^2$, except the zeroth, are equal to zero. Hence $|g|$ is almost everywhere equal to a constant c on $\partial\mathbb{D}$, and since $g \neq 0$, $c \neq 0$.

Since f vanishes on E, so does every element of C. It looks plausible to deduce that all functions in clos C must also vanish on E, but since L^2 convergence does not imply pointwise convergence, we had better justify this carefully. Define a function h on $\partial\mathbb{D}$ by

$$h(z) = \begin{cases} 0 & \text{if } z \in \partial\mathbb{D} \backslash E, \\ |g(z)|/\overline{g(z)} & \text{if } z \in E. \end{cases}$$

Then $h \in L^2$ and $(F, h) = 0$ for all $F \in C$. It follows by continuity that

$(F, h) = 0$ for all $F \in \mathrm{clos}\, C$, and so in particular for $F = g$. That is,

$$0 = (g, h) = \frac{1}{2\pi} \int_E |g(e^{i\theta})|\, d\theta$$

$$= c \times \text{measure of } E \neq 0,$$

a contradiction. Hence f must be the zero function. $\qquad\square$

This elegant proof of a non-trivial function-theoretic property with the aid of simple Hilbert space geometry is taken from Helson (1964), a beautifully written slender volume which digs much deeper in the same vein.

13.4   Multiplication operators and infinite Toeplitz and Hankel matrices

One of the ways functions connect with operators is through multiplication: for a fixed measurable function φ the mapping $f \to \varphi f$ will be a linear operator between suitable function spaces. In Chapter 15 we shall actually construct a function φ on $\partial \mathbb{D}$ by generating the corresponding multiplication operator a little at a time, and so we shall need some observations about such operators and their matrices.

13.14 Theorem   Let φ be a Lebesgue measurable function on $\partial \mathbb{D}$. The following are equivalent:

(i) $\varphi f \in L^2$ for all $f \in L^2$;

(ii) $\varphi \in L^\infty$.

When these conditions hold the operator M_φ on L^2 defined by

$$M_\varphi f = \varphi f, \qquad f \in L^2$$

has norm

$$\| M_\varphi \| = \| \varphi \|_\infty.$$

Proof. (ii) $\Rightarrow$ (i) is easy. To prove the converse, suppose that (i) holds but that $\varphi \notin L^\infty$. This means that, for every $c \in \mathbb{R}$,

$$F(c) = \{ z \in \partial \mathbb{D} : |\varphi(z)| \geq c \}$$

has positive measure in $\partial \mathbb{D}$. Let $c_1 = 1$ and choose $c_2 \geq 2$ such that

$$E_1 = F(c_1) \backslash F(c_2)$$

has non-zero measure. Choose $c_3 \geq 3$ so that

$$E_2 = F(c_2) \backslash F(c_3)$$

has positive measure. Continuing in this way we obtain an infinite sequence $(E_n)_{n \in \mathbb{N}}$ of sets of positive measure which are pairwise disjoint and satisfy

$|\varphi(z)| \geqslant n$ for $z \in E_n$. Let

$$f_n(z) = \begin{cases} m(E_n)^{-1/2} & \text{if } z \in E_n \\ 0 & z \in \partial \mathbb{D} \setminus E_n, \end{cases}$$

where $m(\cdot)$ is Lebesgue measure. Then (f_n) is an orthonormal sequence in L^2, and hence

$$f = \sum_{n=1}^{\infty} n^{-1} f_n$$

exists in L^2. And, for any $N \in \mathbb{N}$, since $f_n \bar{f}_k = 0$ when $n \neq k$,

$$\int |\varphi f|^2 \, dm \geqslant \int |\varphi|^2 \left| \sum_{1}^{N} n^{-1} f_n \right|^2 dm$$

$$= \sum_{1}^{N} \int_{E_n} |\varphi|^2 n^{-2} m(E_n)^{-1} \, dm$$

$$\geqslant \sum_{1}^{N} 1 = N.$$

This contradicts property (i), and hence $\varphi \in L^{\infty}$. Thus (i) and (ii) are equivalent.

Now suppose (i) and (ii) are satisfied. For any $f \in L^2$ we have

$$\|M_\varphi f\|^2 = \int |\varphi f|^2 \, dm \leqslant \|\varphi\|_{\infty}^2 \int |f|^2 \, dm$$

$$= \|\varphi\|_{\infty}^2 \|f\|^2.$$

Hence M_φ is bounded and $\|M_\varphi\| \leqslant \|\varphi\|_{\infty}$. Suppose $\|\varphi\|_{\infty} > \|M_\varphi\|$: then the set

$$\{z : |\varphi(z)| > \|M_\varphi\|\} = \bigcup_{n=1}^{\infty} \{z : |\varphi(z)| \geqslant \|M_\varphi\| + 1/n\}$$

has positive measure, and so one of the summands on the right hand side has positive measure. Thus there exists $\varepsilon > 0$ and a set $E \subseteq \partial \mathbb{D}$ of positive measure such that

$$|\varphi(z)| \geqslant \|M_\varphi\| + \varepsilon, \quad \text{all } z \in E.$$

Let

$$f(z) = \begin{cases} m(E)^{-1/2} & \text{if } z \in E, \\ 0 & \text{if } z \in \partial \mathbb{D} \setminus E. \end{cases}$$

Then $\|f\| = 1$ in L^2 and

$$\|M_\varphi f\|^2 = \int_E |\varphi|^2 m(E)^{-1} \, dm$$

$$\geqslant (\|M_\varphi\| + \varepsilon)^2.$$

This is a contradiction, and so $\|\varphi\|_{\infty} = \|M_\varphi\|$. $\qquad \square$

The matrix of M_φ with respect to the standard orthonormal basis $(z^n)_{-\infty}^\infty$ has an especially regular form. According to Definition 7.23 the column with index j of the matrix consists of the components with respect to $(z^n)_{-\infty}^\infty$ of $M_\varphi z^j$. Now if $\varphi \in L^\infty$ has Fourier series

$$\varphi(z) \sim \sum_{-\infty}^\infty a_n z^n$$

then

$$M_\varphi z^j = z^j \varphi(z) \sim \sum_{-\infty}^\infty a_n z^{n+j}.$$

The jth column of the matrix of M_φ is thus

$$\begin{bmatrix} \vdots \\ a_{-j-1} \\ a_{-j} \\ a_{-j+1} \\ \vdots \end{bmatrix} \quad \begin{matrix} (z^{-1}) \\ (z^0) \\ (z^1) \end{matrix}$$

Since the matrix of M_φ has its rows and its columns indexed by $\mathbb{Z}$ it has no border, and we must indicate in some way which rows and columns correspond to which basis elements. This can be done by marking the entry in row 0, column 0 – say by enclosing it in a square. With this convention the desired matrix is

$$M_\varphi \sim \begin{bmatrix} \cdots & \cdot & \cdot & \cdot & \cdots \\ \cdots & a_0 & a_{-1} & a_{-2} & \cdots \\ \cdots & a_1 & \boxed{a_0} & a_{-1} & \cdots \\ \cdots & a_2 & a_1 & a_0 & \cdots \\ \cdots & \cdot & \cdot & \cdot & \cdots \end{bmatrix}$$

The matrix has constant entries on diagonals $i - j = \text{constant}$. A matrix with this property is called a *Toeplitz matrix*.

In calculating the matrix of

$$M_\varphi : L^2 \to L^2$$

we are not obliged to use the same orthonormal basis in the domain and codomain. Let us consider matrices of multiplication operators with respect to the basis $(z^j)_{-\infty}^\infty$ in the domain and $(z^{-i})_{-\infty}^\infty$ in the codomain. This may seem a perverse thing to do, but the point will become clear in Chapter 15.

13.15 Theorem   The matrix

$$T = [a_{-i-j}]_{i,j=-\infty}^\infty$$

defines a bounded linear operator on $\ell_{\mathbb{Z}}^2$ if and only if there exists $\varphi \in L^\infty$

such that $\hat{\varphi}(n) = a_n$ for all $n \in \mathbb{Z}$. When this holds, T is the matrix of M_φ with respect to the bases $(z^j)^\infty_{-\infty}$ in the domain, $(z^{-i})^\infty_{-\infty}$ in the codomain, and

$$\|T\| = \|\varphi\|_\infty.$$

Proof. If such a φ exists then T is its matrix: this is the calculation we have just carried out, except that the order of the rows is reversed. It follows that $\|T\| = \|M_\varphi\|$, and so by Theorem 13.14,

$$\|T\| = \|\varphi\|_\infty.$$

Now suppose that T is bounded on $\ell^2_{\mathbb{Z}}$. In particular T maps any standard basis vector into $\ell^2_{\mathbb{Z}}$, which is to say that any column of T is square summable. Hence, by Theorem 4.11,

$$\varphi(z) = \sum_{-\infty}^{\infty} a_n z^n \qquad (13.20)$$

exists in L^2. Consider any $f \in L^2$, with Fourier series

$$f(z) = \sum_{-\infty}^{\infty} c_n z^n.$$

By the Cauchy–Schwarz inequality, $|\varphi f|$ is integrable on $\partial \mathbb{D}$, so that the Fourier series of φf exists, and is given by

$$(\varphi f)(z) = \sum_{n=-\infty}^{\infty} \left(\sum_{m=-\infty}^{\infty} a_{n-m} c_m \right) z^n$$

$$= \sum_{i=-\infty}^{\infty} \left(\sum_{j=-\infty}^{\infty} a_{-i-j} c_j \right) z^{-i}$$

(cf. Problem 13.9 for a full argument). Comparing with the action of T on $\ell^2_{\mathbb{Z}}$, we see that, if L^2 functions are identified with their sequences of Fourier coefficients, multiplication by φ corresponds to the application of T. It follows that $\varphi f \in L^2$ for all $f \in L^2$. Thus, by Theorem 13.14, $\varphi \in L^\infty$, while $\hat{\varphi}(n) = a_n$, $n \in \mathbb{N}$, by (13.20). A matrix of the form of T is called a *Hankel matrix*. $\square$

13.5   Problems

13.1. Let $f(z) = \log(1 - z)$, $z \in \partial\mathbb{D} \setminus \{1\}$, where log denotes the branch of the logarithm analytic on the complement of the negative real axis and real on the positive real axis. Show that if $z = e^{i\theta}$, $0 < \theta < 2\pi$, then $f(z) = \log(2 \sin \theta/2) + \frac{1}{2}(\theta - \pi)i$. Hence show that $f \in L^2$.

13.2. Let f be as in Problem 13.1. By integrating $f(z)z^{n-1}$, $n \in \mathbb{N}$, round the unit circle indented at $z = 1$, show that $f \in H^2$. Deduce that H^∞ is a proper subset of H^2.

13.3. Let $f(z) = (z - 1)^{-1}$, $z \in \partial\mathbb{D} \setminus \{1\}$. Prove that $f \notin L^2$.

13.4. Prove that a rational function which belongs to L^2 has no pole on the unit circle, and hence belongs to L^∞.

13.5. Let $\alpha_j \in \mathbb{D}\setminus\{0\}$ for $j \in \mathbb{N}$ and let

$$B_n(z) = \lambda \prod_{j=1}^{n} \frac{|\alpha_j|}{\alpha_j} \frac{\alpha_j - z}{1 - \bar{\alpha}_j z}, \qquad z \in \mathbb{D},$$

where $|\lambda| = 1$. Show that B_n is an inner function (called a *finite Blaschke product*). Show further that, if $n > m$, then

$$\|B_n - B_m\|^2 = 2\left(1 - \prod_{j=m+1}^{n} |\alpha_j|\right)$$

in H^2. Deduce that (B_n) converges in H^2 provided the infinite sum $\sum_{1}^{\infty}(1 - |\alpha_j|)$ is convergent.

13.6. Let B_n be as in Problem 13.5 and suppose $\prod_{1}^{\infty} |\alpha_j|$ converges. Let B be the limit of (B_n) in H^2. Show that if B is not inner then there exists $\varepsilon > 0$ and $E \subseteq \partial\mathbb{D}$ of positive measure $m(E)$ such that $\big||B(z)| - 1\big| \geqslant \varepsilon$ for $z \in E$. Deduce that

$$\|B - B_n\| \geqslant \varepsilon m(E)^{1/2}, \qquad \text{all } n \in \mathbb{N},$$

and conclude that B is inner.

13.7. Construct an inner function φ such that $\lim_{r \to 1-} \varphi(rz)$ fails to exist for some $z \in \partial\mathbb{D}$.

13.8. Let

$$K_r(t) = \frac{4(1 - r^2)\sin^2 t}{(1 - 2r\cos t + r^2)^2}, \qquad 0 \leqslant r < 1, t \in \mathbb{R}$$

(cf. (13.15), (13.16)). Show by integration by parts that

$$\int_0^\pi K_r(t)\,dt = \frac{1}{r}\int_{-\pi}^\pi \cos t\, P_r(t)\,dt,$$

where $0 < r < 1$ and P_r is Poisson's kernel. Deduce that

$$\int_0^\pi K_r(t)\,dt = \frac{1 - r^2}{2ri}\int_{\partial\mathbb{D}} \frac{z^2 + 1}{(1 - rz)(z - r)}\,\frac{dz}{z},$$

and conclude from the residue theorem that

$$\int_0^\pi K_r(t)\,dt = 2\pi.$$

13.9. Let $\varphi, f \in L^2$ have Fourier series

$$\varphi(z) = \sum_{-\infty}^{\infty} a_n z^n, \qquad f(z) = \sum_{-\infty}^{\infty} c_n z^n.$$

Let $\varphi_m(z) = \sum_{n=-m}^{m} a_n z^n$. Show that

$$\frac{1}{2\pi}\int_{-\pi}^\pi \varphi_m(e^{i\theta}) f(e^{i\theta})\, e^{-in\theta}\,d\theta = \sum_{j=-m}^{m} a_j c_{n-j}.$$

Show further that

$$\int_{-\pi}^{\pi} |f(e^{i\theta})||\varphi(e^{i\theta}) - \varphi_m(e^{i\theta})| \, d\theta \to 0$$

as $m \to \infty$, and deduce that

$$\int_{-\pi}^{\pi} \varphi_m(e^{i\theta}) f(e^{i\theta}) \, e^{-in\theta} \, d\theta \to \int_{-\pi}^{\pi} \varphi(e^{i\theta}) f(e^{i\theta}) \, e^{-in\theta} \, d\theta$$

as $m \to \infty$. Conclude that the nth Fourier coefficient of φf is $\sum_{j=-\infty}^{\infty} a_j c_{n-j}$.

13.10. Do Problems 7.5, 7.6, 7.11, 7.13 with RH^2 replaced by H^2.

13.11. Make another attempt at Problem 7.42.

13.12. Let the polynomial

$$p(z) = a_0 + a_1 z + \cdots + a_{n-1} z^{n-1} + a_n z^n,$$

where $a_n \neq 0$, have all its zeros inside $\mathbb{D}$, and let

$$\tilde{p}(z) = \bar{a}_n + \bar{a}_{n-1} z + \cdots + \bar{a}_1 z^{n-1} + \bar{a}_0 z^n.$$

Prove that $\overline{p(z)} = \bar{z}^n \tilde{p}(z)$ when $z \in \partial \mathbb{D}$, and that $1/\tilde{p} \in H^\infty$. Denote by pH^2 the subspace $\{pf : f \in H^2\}$ of H^2. Show that if $f \in H^2 \ominus pH^2$ then $\tilde{p}f \perp z^n H^2$. Deduce that $H^2 \ominus pH^2$ is the space of rational functions expressible in the form $g/\tilde{p}$ where g is a polynomial of degree less than n.

13.13. Let $\alpha_1, \ldots, \alpha_n \in \mathbb{D}$ and let

$$p(z) = (z - \alpha_1)(z - \alpha_2) \ldots (z - \alpha_n).$$

Let $pH^\infty = \{pf : f \in H^\infty\}$. Show that pH^∞ is the space of all bounded analytic functions in $\mathbb{D}$ which vanish at $\alpha_1, \ldots, \alpha_n$. If $\tilde{p}$ is defined as in Problem 13.12 and $\theta = p/\tilde{p}$, show that θ is inner and that $\theta H^\infty = pH^\infty$.

13.14. Let $\alpha_1, \ldots, \alpha_n \in \mathbb{D}$ and let $w_1, \ldots, w_n \in \mathbb{C}$. Say that $f \in H^\infty$ is an *interpolating function* if $f(\alpha_j) = w_j$, $1 \leq j \leq n$. The *Nevanlinna–Pick problem* is to find an interpolating function of minimal norm in H^∞. The purpose of this exercise is to obtain a re-formulation of the problem.

Let p, $\tilde{p}$ and θ be as in Problem 13.13. Choose a polynomial g such that $g(\alpha_j) = w_j$, $1 \leq j \leq n$. Show that the set of interpolating functions coincides with the coset $g + pH^\infty$ in H^∞. Deduce that the infimum of the H^∞-norms of all interpolating functions is equal to

$$\inf\{\|g/\theta - h\|_\infty : h \in H^\infty\}.$$

Show further that if this infimum is attained at $h_0 \in H^\infty$ then an interpolating function of minimal norm is $g - \theta h_0$.

14

Interlude: complex analysis and operators in engineering

Let us pause a while from the technicalities of spaces and operators and reflect on the place of functional analysis in the wider picture. On the strength of remarks earlier in the book about the genesis of the subject we should certainly expect that functional analysis would provide a framework in which to formulate, discuss and (sometimes) solve problems in the description of phenomena on the basis of classical physics. The hanging chain example of Chapters 9–11 is intended to illustrate this aspect. But this is far from being the only role of the subject. It often happens that concepts introduced and developed for one purpose subsequently provide the key to understanding a quite different set of problems. Basic Hilbert space theory is at once close enough to our geometric intuition to be readily assimilated and advanced enough to provide powerful tools for deepening our knowledge over a wide area of mathematics. It is routinely used in differential geometry, complex analysis, number theory – indeed, almost every branch. As we learn more mathematics we come to appreciate better the inter-relationship between its parts, but in view of the brevity of life no individual can fully grasp all the ways in which a particular intellectual current flows through the body of mathematics and on into science. We can still gain some understanding of the process by studying particular instances, as we form conceptions of the natural and social orders from the relatively few places, things and people we know. Chapters 12–16 describe an application of functional analysis to complex function theory. The problem it solves is a natural one, of independent interest: that is, it is a problem which would (and did) occur to complex analysts working without the perspective of functional analysis. It is even a fundamental problem, in the sense that it often recurs as a step in attacks on more advanced problems. The feature of this problem which makes it especially suitable as an illustration of the workings of mathematics is that it has direct application in engineering: it is one of the theoretical components in one approach to the design of certain engineering devices, notably automatic controllers. Those who like their mathematics self-contained and are not inspired by its relation to the physical world may proceed at once to the next chapter.

Gadgets which control machines automatically are all round us and are becoming ever more significant to our society. Their proliferation is usually reckoned to have started around 1788, with the widespread introduction of James Watt's governor. This was a simple mechanical device for regulating the speed of steam engines. Steel balls are suspended on rods attached to a vertical axle driven by the engine. As the angular velocity of the axle increases, the balls rise and this movement is communicated through levers to a valve which reduces the supply of steam to the engine and so restores the angular velocity to a lower value. Contrary to myth, Watt was not the originator of this idea; nor did he contribute to its theory. His success was evidently due to good production engineering: he tried out numerous ways of regulating engines, and was ultimately able to market a governor which worked reliably.

Although Watt's governor was adequate for the requirements of industrial steam engines, the technique fell short of perfection. Astronomers particularly had very stringent requirements for the accuracy of the clockwork mechanisms which compensated for the rotation of the earth and kept their telescopes aligned with the stars. They found that in practice governors brought about angular velocities which, instead of being constant, oscillated about the equilibrium value. In unfavourable circumstances these oscillations could dominate the motion. These phenomena were first discussed and treated mathematically by the Astronomer Royal, George Airy, in 1840. Referring to the operation of frictional forces, Airy wrote: 'In inadvertence of their effect I constructed (some time since) a machine for uniform motion. It went exceedingly well till inequality began to be perceptible; this inequality, though still of an oscillatory character, increased rapidly, and finally became so great as entirely to destroy the character of the previous motion; and the machine (if I may so express myself) became perfectly wild.'

Such behaviour is now known more prosaically as dynamic instability: it remains something to be avoided above all else in controlling devices. Airy's mathematical treatment was not definitive: the foundations of a proper treatment of stability were laid by Maxwell in 1868 (claims are also made for Vishnegradskii and Stodola). Although Maxwell established important principles, he certainly did not say the last word. Since then the subject of control has developed an enormous body of experience and

theory, bridging engineering and mathematics. There are a dozen or so
international journals on the topic, printing thousands of papers a year.
There are papers on general principles common to many cases, others on
special features of, say, automatic pilots or the control of chemical
processes. People make use of algebra, differential geometry, Riemann
surfaces, stochastic processes, non-linear dynamics – every major branch of
contemporary mathematics. For illustration consult a recent volume of the
IEEE Transactions on Automatic Control. Operator theory is but one of the
mathematical strands in this tapestry, but lately it has come to play a
substantial role in a promising new approach to one of the central concerns
of control: how to achieve stabilization and acceptable performance in the
face of imperfect knowledge of the system to be controlled. In this chapter I
hope to convey an inkling of how this connection is made by means of a
drastically simplified presentation.

  The first step must be to find a mathematical representation of the
physical system which it is desired to control, be it a ship, a steam engine, a
chemical works or whatever. This stage takes up a high proportion of the
time and effort of the working control engineer, but here we shall assume it
has been done. The jargon word, which does for both the physical system
and its mathematical representation, is the *plant*. We suppose that the state
of the plant at any instant can be described by a finite sequence of real
numbers. For example, in the case of an automatic pilot we might take the
parameters of interest to be the co-ordinates of the centre of gravity and the
Euler angles of the aircraft with respect to suitable co-ordinate axes,
together with the derivatives of these quantities with respect to time. The
state of the system would then be represented by a vector in $\mathbb{R}^{12}$. We
suppose further that we can influence the behaviour of the system by means
of a finite number of 'control inputs', each represented by a real number.
Thus, in the case of an aircraft, we can adjust the thrust of the engines and
the settings of the rudder, elevons, etc. The state of the system and the
control inputs at time t are consequently represented by vectors $x(t) \in X$,
$u(t) \in U$ where X and U are Euclidean spaces of suitable dimension. A
model of the control system consists of a relation between $x(\cdot)$ and $u(\cdot)$,
most commonly a differential equation. In the flight example there are six
basic equations, corresponding to components of force along and torque
about each of the three co-ordinate axes. These equations involve
trigonometric functions of the Euler angles, and so are non-linear.
However, for the analysis of small deviations from an operating
equilibrium we may replace the non-linear terms by linear approximations,
just as one replaces $\sin \theta$ by θ in discussing the small oscillations of a simple
pendulum. Control is thus usually based on modelling the plant by a linear

differential equation, though typically several different linear models will be needed for a single plant: clearly different linear approximations will be appropriate to an aeroplane in steady flight and during landing.

Let us consider a plant modelled by the equations

$$Lx = Mu,$$

where L, M are linear differential operators, x and u being functions of time with values in Euclidean spaces X and U. It is natural to assume that the physical characteristics of the plant do not change over the period during which control is to be exercised, and this corresponds to assuming that L, M are linear differential operators with constant coefficients. On taking Laplace transforms we obtain (assuming, for simplicity, zero initial conditions)

$$A(s)\bar{x}(s) = B(s)\bar{u}(s) \tag{14.1}$$

where $\bar{x}(s)$, $\bar{u}(s)$ are the Laplace transforms of $x(t)$, $u(t)$ and $A(s)$, $B(s)$ are matrix-valued polynomials in s. An adequate model must contain enough information to determine $x(\cdot)$ from $u(\cdot)$, so we may take it that (14.1) may be solved to give

$$\bar{x}(s) = G(s)\bar{u}(s), \tag{14.2}$$

where $G(s)$ is a matrix of rational functions in s. $G(s)$ is called the *transfer function matrix*, or simply *transfer function*, of the system.

What an encouraging conclusion for the pure mathematician! Such a marvel of engineering as a modern aircraft, with all its complexity and power, can be usefully represented by a few rational matrix functions. True, they may be rather large; a lot of information is doubtless encoded in their coefficients and someone may have to do a lot of work to estimate them, but in principle, with the great structures of linear algebra, complex analysis, ring theory and functional analysis to support us, we should feel at home with them.

We have seen that an undesirable feature of a physical device is instability. We must translate this into a statement about transfer functions. Consider a system whose state at time t can be represented by a single real number $x(t)$, with a single control input $u(t)$, governed by the equations

$$\ddot{x} + \dot{x} - 6x = u(t), \tag{14.3}$$

$$x(0) = \dot{x}(0) = 0.$$

The transfer function of this plant is

$$G(s) = \frac{1}{s^2 + s - 6} = \frac{1}{(s-2)(s+3)},$$

and the equations can be solved to give

$$x(t) = Ae^{2t} + Be^{-3t} + (h * u)(t),\qquad(14.4)$$

where $h * u$ is a particular integral of (14.3) and A, B are constants. It should be easy to accept that the term Ae^{2t} in (14.4) would be troublesome for a real system: presumably we do not want the state to go off to infinity as time passes. This term is present because the transfer function G has a pole at $s = 2$. Any pole of G in the right half plane $\{s \in \mathbb{C} : \mathrm{Re}\, s > 0\}$ will give rise to an exponential term in the solution for $x(\cdot)$ which tends to infinity with t. One of Maxwell's contributions was to identify stable systems as those whose transfer functions are bounded and analytic in the right half plane (actually the issue is a good deal more subtle, but this will do for present purposes).

One of the main points of the use of automatic controllers is to get an initially unstable system to behave stably. In the case of the system described by equation (14.4) this entails choosing $u(\cdot)$ so that the particular integral term $h * u$ counterbalances the destabilizing exponential term Ae^{2t}. This can be achieved by means of a 'feedback loop', which we can describe with the aid of a block diagram. The uncontrolled plant with transfer function G is represented by Diagram 14.1. This is simply another way of writing the equation $x = Gu$, which is equation (14.2) but for the fact that we are dispensing with the bars which indicate Laplace transforms. A feedback controller takes some measurement of the state function $x(\cdot)$, transforms it in some way and subtracts the result from the input to the system. If a human operator inputs a signal $u(t)$ and the state at time t is $x(t)$, then the input which actually reaches the system is $u(t) - (Kx)(t)$, where K is some operator which represents the action of the controlling device. This set-up can be shown by Diagram 14.2.

The engineer will naturally design the controller so that its action can be analysed, and this usually means that he chooses a controller which can be modelled by linear constant coefficient differential equations. Thus the controller will also be represented by a transfer function, and so we may regard all of u, x, G and K as functions of the complex variable s, of suitable (vector or matrix) types. Since the plant has input $u - Kx$ and state x we have

$$x = G(u - Kx)$$

Diagram 14.1

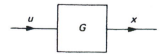

and hence

$$x = (I + GK)^{-1}Gu.$$

This shows that the compound plant, consisting of the original plant together with the feedback controller K in the configuration of Diagram 14.2 is

$$(I + GK)^{-1}G.$$

To take an example of extreme simplicity, the function

$$G(s) = \frac{1}{s - 3}$$

represents an unstable system, having a pole at $s = 3$. If we choose $K(s)$ to be constant and equal to 4, we find

$$(I + GK)^{-1}G(s) = \frac{1}{s + 1},$$

and so the feedback controller K stabilizes the plant.

Returning once again to Diagram 14.2 we observe that the input which actually reaches the plant is

$$\begin{aligned} u - Kx &= u - K(I + GK)^{-1}Gu \\ &= u - (I + KG)^{-1}KGu \\ &= (I + KG)^{-1}u. \end{aligned}$$

If it happens that the transfer function $(I + KG)^{-1}$ has a pole in the right half plane then the compound plant will have instabilities inside it, even though its overall transfer function may be free of right half plane poles. Similar considerations lead us to say that the feedback system of Diagram 14.2 is *internally stable* if $(I + GK)^{-1}$, $(I + GK)^{-1}G$, $K(I + GK)^{-1}$ and $(I + KG)^{-1}$ are all analytic in the right half plane (they are allowed to have removable singularities). Suppose the plant G is given to us and we seek a

Diagram 14.2

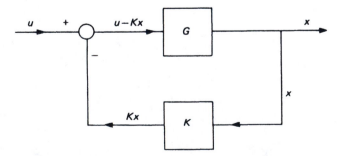

description of all controllers K which will make the feedback system internally stable. This problem has a very elegant solution: it is elementary but far from obvious. Starting from G one can construct polynomial matrices A, B, C, D of appropriate types so that the desired controllers K are precisely those expressible in the form

$$K = (AQ + B)(CQ + D)^{-1} \qquad (14.5)$$

for some rational matrix Q analytic in the right half plane. This is called the *Youla parametrization* (see Francis, 1986, Chapter 4). If it were just a matter of finding *some* internally stabilizing controller we could simply take $Q = 0$. In a real design problem, however, there will be many other factors to be taken into consideration. How well will the system stand up to unpredictable external disturbances (gusts of wind or a stewardess wheeling a drinks trolley down the aisle)? What will be the effect of the infinitesimal delays which are inevitable in a real physical construct? Presumably the exercise of control costs something: can we avoid undue extravagance? Clearly there is room for cunning in the choice of Q. This is no place to tackle all the complexities of a real engineering system: the purpose of this chapter will be served if we can appreciate how functional analysis enters into the treatment of one important problem: robust stabilization.

A mathematical model never describes the behaviour of a plant exactly. A controller which looks good on paper may turn out in practice to be so sensitive to small variations in model parameters or to deviations of the real system from the model (such as non-linearities) as to be useless. Some design methods are better than others in this respect, and engineers rely on their experience and intuition to select a design and on thorough testing to check its performance. The theory of feedback has been a relatively small part of the practice of control engineers. However, the march of technology is making for new requirements. As control systems have to perform more and more complicated tasks, intuition becomes harder to attain. It is difficult to develop a feel for a system with many inputs and outputs. Furthermore, it can happen that testing is impractical: a satellite which is designed to operate in zero gravity cannot be tested at the surface of the earth and has to work first time. To some extent testing the real thing can be replaced by computer simulation, but there is a clear need for a theory which takes account of the imprecise nature of mathematical models. This brings us to the notion of robustness. A design is *robust* if it works not only for the postulated model, but also for neighbouring models. A recent approach (since 1980) to formalizing this notion is to interpret closeness of models as closeness of their transfer functions in the H^∞-norm. There are good physical grounds for this criterion. Indeed, one might wonder why this idea was such a late starter. Part of the answer must be that engineers were unaware of the relevant mathematical theorems and operator theorists of the engineering problem. Credit for making the connection seems to be due to G. Zames and J. W. Helton (see references in Francis, 1986). The theory is now developing rapidly and looks to be a promising new tool.

What happens if we try to add a dash of robustness to the problem of internal stabilization? Suppose we are given a transfer function G, but suspect it may be in error by up to ε. We believe that the 'real' transfer function of the system is $G + \Delta G$, where $G + \Delta G$ has the same number of poles in the closed right half plane as G and

$$\|\Delta G(s)\| < \varepsilon, \qquad \text{all Re } s > 0$$

(here $\|\cdot\|$ denotes the operator norm on $\mathscr{L}(U, X)$). Let us denote the class of such functions $G + \Delta G$ by $V(G, \varepsilon)$. Since we do not know ΔG we need to find a controller which stabilizes everything in $V(G, \varepsilon)$. It is not hard to show that a rational function K internally stabilizes every function in $V(G, \varepsilon)$ if and only if

(i) K internally stabilizes G, and
(ii) $\|K(I - GK)^{-1}(s)\| \leqslant 1/\varepsilon$, all Re $s > 0$.

There may or may not exist a K which satisfies these two conditions: it depends on the characteristics of G and the size of ε. To ascertain whether there does we take from (14.5) the parametrization of all K which satisfy (i) and substitute in (ii). By virtue of some happy coincidences the result simplifies better than we are entitled to expect, and we find that

$$K(I - GK)^{-1} = T_1 + T_2 Q T_3$$

where T_1, T_2 and T_3 are polynomial matrices and Q is the free parameter (a rational function analytic and bounded on the right half plane).

In summary, if we are given a transfer function G and a tolerance ε, we can determine whether there exists a controller (representable by a rational matrix function) which internally stabilizes all plants lying within ε of G in the H^∞-norm and having the same number of 'unstable poles' as G provided we can solve a certain concrete problem about analytic matrix functions. That is, we can constructively find polynomial matrices T_1, T_2 and T_3, derived from G, such that the answer to the robust stabilization question is yes or no depending on whether there does or does not exist a bounded analytic matrix function Q in the right half plane such that

$$\|(T_1 + T_2 Q T_3)(s)\| \leqslant 1/\varepsilon, \qquad \text{all Re } s > 0.$$

Furthermore, if we can find such a Q, then on substituting it in (14.5) we shall obtain the desired robustly stabilizing controller.

Determining whether there is such a Q (and finding one if there is) is called by engineers the *model matching problem*. For the special case of scalar functions it was solved long ago. The classical formulation was as an interpolation problem. Suppose that

$$T_2(s)\, T_3(s) = \prod_{j=1}^{n} s - s_j,$$

where $s_1, \ldots, s_n$ are distinct points in the right half plane. Let us for the moment denote by H^∞ the space of functions bounded and analytic in the right half plane (this conflicts slightly with the usage in other chapters, but since the half-plane and disc are related by the Cayley transform $z = (s - 1)/(s + 1)$, there is no essential difference). The class of functions expressible as $T_1 + T_2 Q T_3$, for some $Q \in H^\infty$, consists precisely of the functions $F \in H^\infty$ satisfying $F(s_j) = T_1(s_j)$, $1 \leqslant j \leqslant n$. Hence the problem reduces to ascertaining whether there exists a function $F \in H^\infty$ meeting a finite number of interpolation conditions and with supremum norm no greater than $1/\varepsilon$. This is known as the Nevanlinna–Pick problem (see Problem 13.14). Numerous solutions of this problem have been put forward, from around 1917 onwards, many of them either implicitly or explicitly operator-theoretic in character. One method is given in Theorems 15.14 and 15.16. In simple cases solutions can be found by hand calculation (see Problems 15.12, 15.13), but usually computers are required.

The classical results of complex analysis are often not quite what the engineers want. In particular, transition from scalar to matrix-valued functions often involves substantial new difficulties, and an operator-theoretic viewpoint seems to be a definite advantage in coping with these. Operator theorists have studied matrix versions of problems analogous to Nevanlinna–Pick since the 1950s, but the mathematical literature in this area is relatively thin, and much of the detailed work has been done only recently by engineers. Several groups have devised and implemented effective numerical algorithms for solving the model matching problem with matrix-valued functions.

There is a danger that this chapter will leave the impression that the problems of control system design are essentially solved, at least as far as the H^∞ approach goes. This is false. The foregoing account deals only with robust stabilization: a realistic design problem would entail many other considerations, which would lead to difficult mathematical problems. Both theoretical and practical aspects of the H^∞ approach are being intensively studied.

15

Approximation by analytic functions

A problem which commonly arises in statistics, physics and engineering is to find the function in a given class which approximates some known function as well as possible. To turn this vague notion into a well-posed mathematical question we need a numerical criterion of how good an approximation is. A popular choice is to look for a best 'least squares' fit. That is, we ask that the norm of the error function (the difference between the given function and the chosen approximation) be as small as possible, where the norm used is a Hilbert space norm. The advantage of this type of criterion is the obliging nature of Hilbert space geometry: such problems are usually comparatively easy to solve. Let us check this in a particular case.

15.1 Problem Given a function $\varphi \in L^2$, find a function $g \in H^2$ such that $\|\varphi - g\|_{L^2}$ is minimized. $\qquad\qquad\qquad\qquad\qquad\qquad\qquad\qquad\qquad\Box$

If we were to write out the definition of $\|\varphi - g\|$ the reason for the expression 'least squares' would be apparent. In this instance the class of functions from which the approximation is to be chosen is a closed subspace of L^2, and in this case the best approximation problem was essentially solved in Chapter 4. Let us recall the relevant facts and introduce some notation. If M is a closed subspace of a Hilbert space H and $x \in H$ then, by Theorem 4.24, there exist $y \in M$, $z \in M^\perp$ such that $x = y + z$. Furthermore, y and z are uniquely determined by these conditions, for if $x = y' + z'$, with $y' \in M$, $z' \in M^\perp$ then

$$y - y' = z' - z \in M \cap M^\perp = \{0\}.$$

The following definition therefore makes sense.

15.2 Definition   Let M be a closed subspace of a Hilbert space H. The *orthogonal projection* from H to M is the operator $P: H \to M$ defined by

$$Px = y$$

if $x = y + z$, where $y \in M$, $z \in M^{\perp}$.   □

It is routine to check that P is linear and that $\|P\| \leqslant 1$.

15.3 Theorem   Let M be a closed subspace of a Hilbert space H and let $x \in H$. The element y of M for which $\|x - y\|$ is minimized is given by $y = Px$, where P is the orthogonal projection from H to M.

Proof. This is the trivial half of Lemma 4.23.   □

15.4 Corollary (Solution of Problem 15.1)   Let $\varphi \in L^2$. The function $g \in H^2$ for which $\|\varphi - g\|_{L^2}$ is minimized over H^2 is

$$g = P_+ \varphi$$

where

$$P_+ : L^2 \to H^2$$

is the orthogonal projection operator.   □

This really is a solution: it enables us to calculate best approximations in H^2, as the following show.

15.5 Examples   (i) Let

$$\varphi(z) = \frac{z+1}{(z - \frac{1}{2})(z + 2)}, \qquad z \in \partial \mathbb{D}.$$

We wish to find $g \in H^2$ which approximates φ as closely as possible in the L^2 norm. By the corollary, it is given by $g = P_+ \varphi$. To calculate g resolve φ into partial fractions:

$$\varphi(z) = \frac{3}{5(z - \frac{1}{2})} + \frac{2}{5(z + 2)}.$$

The second term on the right hand side is bounded and analytic on the closed unit disc. By Theorem 13.11 it belongs to H^∞, and so *a fortiori* it is in H^2. And

$$\frac{3}{5(z - \frac{1}{2})} = \frac{3}{5}\left(\frac{1}{z} + \frac{1}{2z^2} + \frac{1}{2^2 z^3} + \cdots \right)$$

uniformly for $|z| \geqslant c > \frac{1}{2}$. Thus $3/5(z - \frac{1}{2})$ is in the closed linear span in L^2 of the functions $(z^n)_{n=-\infty}^{-1}$, which is $(H^2)^{\perp}$. Thus

$$(P_+ \varphi)(z) = \frac{2}{5(z + 2)}.$$

It is clear that this method will work for any rational $\varphi \in L^2$: $P_+\varphi$ is the sum of all those partial fractions of φ whose poles lie outside $\mathbb{D}$. See Problem 15.4.

(ii) Let

$$\varphi(e^{i\theta}) = -i(\pi - \theta), \qquad 0 \leqslant \theta < 2\pi.$$

How do we find $P_+\varphi$ here? Calculate the Fourier series of φ – say

$$\varphi(z) = \sum_{-\infty}^{\infty} a_n z^n,$$

the equality being of course in the sense of L^2 convergence. We have

$$\sum_{-\infty}^{\infty} a_n z^n = \sum_{-\infty}^{-1} a_n z^n + \sum_{0}^{\infty} a_n z^n.$$

The two terms on the right hand side clearly belong to $(H^2)^{\perp}$ and H^2 respectively. Hence

$$P_+\left(\sum_{-\infty}^{\infty} a_n z^n \right) = \sum_{0}^{\infty} a_n z^n.$$

Calculation shows that in the present example

$$a_n = \begin{cases} 0 & \text{if } n = 0 \\ -\dfrac{1}{n} & \text{if } n \neq 0. \end{cases}$$

Thus

$$(P_+\varphi)(z) = -\sum_{n=1}^{\infty} \frac{z^n}{n}$$

$$= \log(1 - z).$$

Hence the best approximation to φ in the L^2-norm by an H^2 function is the function $\log(1 - z)$. $\qquad\square$

15.1 The Nehari problem

Best approximation problems with respect to norms which are not Hilbert space norms can hardly ever be solved either so easily or so explicitly. For this reason engineers and others often have to make do with L^2-type criteria even when the physics of their problems call for the minimization of some different norm: there may simply be no known satisfactory way of carrying out the minimization. Fortunately there is one other class of norms which has a clear physical interpretation and is mathematically tractable. Use of the L^∞-norm is natural in many contexts, including those of the examples in the last chapter, and the corresponding best approximation problems can often be solved on a computer, even though the mathematics is considerably less straightforward than for L^2.

Although the L^∞-norm is not a Hilbert space norm it is still closely related to Hilbert space: as in Theorem 13.14, the L^∞-norm of φ is the same as the operator norm of multiplication by φ on L^2. The rich theory of operators on Hilbert space therefore comes to our aid. We shall illustrate how this happens by solving a sample problem.

15.6 *The Nehari problem*   Given $\varphi \in L^\infty$, find a function $g \in H^\infty$ such that $\|\varphi - g\|_\infty$ is minimized.   □

This differs from Problem 15.1 only in that the L^2-norm has been replaced by the L^∞-norm. One would certainly expect changing the norm to change the solution, but we had nevertheless better check that it is so. Consider Example 15.5(ii): $\varphi(e^{i\theta}) = -i(\pi - \theta), 0 < \theta < 2\pi$. This is certainly an L^∞-function: might not its best analytic (i.e. H^∞) approximation with respect to the L^∞-norm be simply $P_+\varphi$, as it was for L^2? The answer is no: as we calculated above, $(P_+\varphi)(z) = -\log(1 - z)$, which is not essentially bounded on $\mathbb{D}$. To get the best approximation we have to use operators.

15.2   Hankel operators

Let

$$P_- : L^2 \to L^2 \ominus H^2$$

be the orthogonal projection operator, so that

$$P_-\left(\sum_{-\infty}^{\infty} a_n z^n\right) = \sum_{-\infty}^{-1} a_n z^n.$$

15.7 *Definitions*   Let $\varphi \in L^\infty$. The *Hankel operator with symbol* φ, denoted by H_φ, is the operator

$$P_- M_\varphi | H^2 : H^2 \to L^2 \ominus H^2,$$

where M_φ is the operation of multiplication by φ on L^2 and '$| H^2$' denotes restriction to H^2.

A *Hankel matrix* is a finite or infinite matrix which is constant on cross-diagonals ($i + j = \text{constant}$).   □

Thus

$$H_\varphi f = P_-(\varphi f), \qquad f \in H^2.$$

A 3×3 Hankel matrix has the form

$$\begin{bmatrix} a_1 & a_2 & a_3 \\ a_2 & a_3 & a_4 \\ a_3 & a_4 & a_5 \end{bmatrix}.$$

15.8 Theorem   Let $\varphi \in L^\infty$ have Fourier series

$$\varphi(z) \sim \sum_{-\infty}^{\infty} a_n z^n.$$

The matrix of H_φ with respect to the bases $1, z, z^2, \ldots$ in H^2 and $z^{-1}, z^{-2}, z^{-3}, \ldots$ in $L^2 \ominus H^2$ is the Hankel matrix

$$\begin{bmatrix} a_{-1} & a_{-2} & a_{-3} & \cdots \\ a_{-2} & a_{-3} & a_{-4} & \cdots \\ \cdot & \cdot & \cdot & \cdots \end{bmatrix}.$$

Proof. The jth column, $j = 0, 1, 2, \ldots$, contains the components of $H_\varphi z^j$ with respect to the orthonormal basis $z^{-1}, z^{-2}, \ldots$ of $L^2 \ominus H^2$. Now

$$H_\varphi z^j = P_-(\varphi z^j)$$

$$= P_-\left(\sum_{n=-\infty}^{\infty} a_n z^{n+j} \right)$$

$$= P_-\left(\sum_{m=-\infty}^{\infty} a_{m-j} z^m \right)$$

$$= \sum_{m=-\infty}^{-1} a_{m-j} z^m$$

$$= a_{-j-1} z^{-1} + a_{-j-2} z^{-2} + a_{-j-3} z^{-3} + \cdots$$

Thus the jth column of the matrix of H_φ is

$$\begin{bmatrix} a_{-j-1} & a_{-j-2} & \cdots \end{bmatrix}^T, \qquad j = 0, 1, 2, \ldots \qquad \square$$

15.9 Example   Hilbert's Hankel matrix. Consider the function

$$\varphi(e^{i\theta}) = -i(\pi - \theta), \qquad 0 \leqslant \theta < 2\pi \tag{15.1}$$

of Examples 15.5(ii). In the notation of the preceding theorem,

$$a_n = -\frac{1}{n}, \qquad n = -1, -2, -3, \ldots.$$

Hence the matrix of the Hankel operator H_φ with respect to the standard bases of H^2 and $L^2 \ominus H^2$ (i.e. the bases described in Theorem 15.8) is

$$\Gamma = \begin{bmatrix} 1 & \frac{1}{2} & \frac{1}{3} & \cdots \\ \frac{1}{2} & \frac{1}{3} & \frac{1}{4} & \cdots \\ \frac{1}{3} & \frac{1}{4} & \frac{1}{5} & \cdots \\ \cdot & \cdot & \cdot & \cdots \end{bmatrix}.$$

Γ is known as Hilbert's Hankel matrix. $\square$

The custom of using 'Γ' to denote a Hankel matrix is due to the influence of Saints Cyril and Methodius. We consider the question: what is the norm of Γ as an operator on ℓ^2? There is a beautiful and surprising answer, to wit

π. If one starts from the matrix Γ itself it is far from obvious that Γ even acts as a bounded operator on ℓ^2. The way we have arrived at Γ here makes it easier to see, since Γ will be bounded if and only if H_φ is bounded, and H_φ is easier to analyse.

If E is a normed space, $A \subseteq E$ and $x \in E$ then $\text{dist}_E(x, A)$ will denote the distance of x from A in E: that is,

$$\text{dist}_E(x, A) = \inf_{y \in A} \|x - y\|_E.$$

15.10 Lemma For any $\varphi \in L^\infty$,

$$\|H_\varphi\| \leqslant \text{dist}_{L^\infty}(\varphi, H^\infty).$$

Proof. For $f \in H^2$ we have

$$\begin{aligned}
\|H_\varphi f\| &= \|P_- M_\varphi f\| \\
&= \|P_-(\varphi f)\| \\
&\leqslant \|\varphi f\| \\
&\leqslant \|\varphi\|_\infty \|f\|,
\end{aligned}$$

the last step by Theorem 13.14. Hence

$$\|H_\varphi\| \leqslant \|\varphi\|_\infty, \tag{15.2}$$

by the definition of the operator norm. Now observe that H_φ depends only on the 'anti-analytic' part of φ, i.e. on $P_-\varphi$. One way of seeing this is from the matrix of H_φ (cf. Theorem 15.8): it depends only on the negative Fourier coefficients of φ, which are determined by $P_-\varphi$. However, it is easy enough to see directly. If $g \in H^\infty$ then, for any $f \in H^2$, $gf \in H^2$ (Theorem 13.12) and hence $P_-(gf) = 0$. Thus

$$\begin{aligned}
H_{\varphi - g} f = P_- M_{\varphi - g} f &= P_-(\varphi f - gf) \\
&= P_-(\varphi f) = H_\varphi f.
\end{aligned}$$

That is,

$$H_{\varphi - g} = H_\varphi, \qquad \text{all } g \in H^\infty.$$

From (15.2)

$$\begin{aligned}
\|H_\varphi\| &= \|H_{\varphi - g}\| \\
&\leqslant \|\varphi - g\|_\infty,
\end{aligned}$$

all $g \in H^\infty$. Taking the infimum of the right hand side over $g \in H^\infty$ gives

$$\|H_\varphi\| \leqslant \text{dist}_{L^\infty}(\varphi, H^\infty). \qquad \square$$

The reason this result is called a lemma and not a theorem is that we shall shortly prove the opposite inequality as well. That, however, is a substantially deeper statement. First let us get an easy picking.

15.11 Corollary   The norm of Hilbert's Hankel matrix, as an operator on ℓ^2, is at most π.

Proof. Since Γ is the matrix of H_φ, where φ is given by (15.1), with respect to orthonormal bases, $\|H_\varphi\| = \|\Gamma\|$. Clearly $\|\varphi\|_\infty = \pi$, and so

$$\|\Gamma\| = \|H_\varphi\| \leqslant \|\varphi\|_\infty = \pi. \qquad \square$$

This result looks even more striking when stated in non-technical language.

15.12 Hilbert's inequality   If $(x_n), (y_n) \in l^2$ then

$$\left| \sum_{n,m=1}^{\infty} \frac{x_n \bar{y}_m}{n+m-1} \right| \leqslant \pi \left\{ \sum_1^\infty |x_n|^2 \right\}^{1/2} \left\{ \sum_1^\infty |y_m|^2 \right\}^{1/2}.$$

Proof. This is a re-statement of the relation

$$|(\Gamma x, y)| \leqslant \pi \|x\| \|y\|$$

which follows from Corollary 15.11 and the Cauchy–Schwarz inequality. $\qquad \square$

We proceed to the opposite inequality of Lemma 15.10. We shall construct a matrix acting on ℓ_Z^2, starting with a south east corner and building it up by successively adding new rows and columns. We need to know that if all the intermediate matrices determine contractions on ℓ^2 then the whole matrix determines a contraction on ℓ_Z^2.

15.13 Lemma   Let T be the infinite matrix

$$T = [t_{ij}]_{i,j=-\infty}^{\infty}$$

and, for $k, l \in \mathbb{Z}$ let T_{kl} be the submatrix of T

$$T_{kl} = [t_{ij}]_{i=k,j=l}^{\infty}.$$

If T_{kl} defines a contraction on ℓ^2 for every $k, l \in \mathbb{Z}$ then T defines a contraction on ℓ_Z^2.

Proof. If M is the matrix of a contraction with respect to orthonormal bases then any column of M consists of the components of the image of a basis vector and so is an ℓ^2 sequence of norm $\leqslant 1$. Applying this observation to M^* we deduce that every row of M is an ℓ^2 sequence of norm $\leqslant 1$. Taking $M = T_{kl}$ we obtain

$$\sum_{j=l}^{\infty} |t_{kj}|^2 \leqslant 1$$

for all $k, l \in \mathbb{Z}$, and hence

$$\sum_{j=-\infty}^{\infty} |t_{kj}|^2 \leqslant 1, \qquad k \in \mathbb{Z}.$$

It follows that if $x = (x_n)_{-\infty}^{\infty} \in \ell_{\mathbb{Z}}^2$, then

$$y_k = \sum_{j=-\infty}^{\infty} t_{kj} x_j$$

exists in $\mathbb{C}$, so we may write

$$Tx = y = (y_k)_{-\infty}^{\infty}.$$

Suppose $\|x\| = 1$ in $\ell_{\mathbb{Z}}^2$. Then

$$x^l = (x_l, x_{l+1}, x_{l+2}, \ldots)$$

has norm at most 1 in ℓ^2, and so $\|T_{kl} x^l\| \leqslant 1$. That is,

$$\sum_{i=k}^{\infty} \left| \sum_{j=l}^{\infty} t_{ij} x_j \right|^2 \leqslant 1.$$

Letting $l \to -\infty$ we obtain

$$\sum_{i=k}^{\infty} |y_i|^2 \leqslant 1, \qquad k \in \mathbb{Z},$$

and then letting $k \to -\infty$ we find

$$\sum_{i=-\infty}^{\infty} |y_i|^2 \leqslant 1.$$

Thus $y \in \ell_{\mathbb{Z}}^2$ and $\|y\| \leqslant 1$. Hence T defines a contraction on $\ell_{\mathbb{Z}}^2$.   □

15.14 Theorem   Let $\varphi \in L^\infty$. Then

$$\|H_\varphi\| = \mathrm{dist}_{L^\infty}(\varphi, H^\infty). \tag{15.3}$$

Furthermore there exists $\psi \in L^\infty$ such that $H_\varphi = H_\psi$ and

$$\|\psi\|_\infty = \|H_\varphi\|.$$

Proof. Lemma 15.10 asserts that

$$\|H_\varphi\| \leqslant \mathrm{dist}_{L^\infty}(\varphi, H^\infty).$$

In proving the opposite inequality we may suppose $\|H_\varphi\| = 1$, replacing φ by $\lambda\varphi$ for some suitable scalar λ. Let $\hat{\varphi}(-n) = a_{-n}, n \in \mathbb{N}$. By Theorem 15.8

$$\Gamma_0 = \begin{bmatrix} a_{-1} & a_{-2} & \cdots \\ a_{-2} & a_{-3} & \cdots \\ \cdot & \cdot & \cdots \end{bmatrix}$$

is the matrix of H_φ with respect to orthonormal bases and hence has norm 1 as an operator on ℓ^2. Construct $a_0, a_1, \ldots \in \mathbb{C}$ inductively as follows. Suppose that $a_j, 0 \leqslant j < k$ are such that the infinite Hankel matrix

$$\Gamma_k = \begin{bmatrix} a_{k-1} & a_{k-2} & \cdots & a_0 & a_{-1} & \cdots \\ a_{k-2} & a_{k-3} & \cdots & a_{-1} & a_{-2} & \cdots \\ \cdot & \cdot & \cdots & \cdot & \cdot & \cdots \end{bmatrix}$$

is a contraction, $k \in \mathbb{Z}^+$. Note that this is so when $k = 0$. We wish to adjoin a new first column so as to preserve the Hankel structure and the property of being a contraction. The Hankel pattern fixes all the new first column except its first entry, which is at our disposal. Let

$$\Gamma_k(p) = \begin{bmatrix} p & a_{k-1} & a_{k-2} & \cdots \\ a_{k-1} & a_{k-2} & a_{k-3} & \cdots \\ \cdot & \cdot & \cdot & \cdots \end{bmatrix}$$

$$= \begin{bmatrix} p & Q \\ R & S \end{bmatrix}, \quad \text{say,}$$

where Q, R and S are suitable matrices of types $1 \times \infty$, $\infty \times 1$ and $\infty \times \infty$ respectively. We have

$$[R \quad S] = \Gamma_k = \begin{bmatrix} Q \\ S \end{bmatrix},$$

and Γ_k is a contraction, by the inductive hypothesis. By Parrott's theorem (12.22) there exists $p \in \mathbb{C}$ such that $\Gamma_k(p)$ is a contraction: let $a_k = p$. Now

$$\Gamma_{k+1} = \Gamma_k(a_k)$$

and is a contraction. By induction the sequence $(a_k)_0^\infty$ has the property that $\|\Gamma_k\| \leqslant 1$ for all $k \geqslant 0$.

Consider the infinite Hankel matrix

$$T = [a_{-i-j+1}]_{i,j=-\infty}^\infty.$$

In the notation of Lemma 15.13

$$T_{kl} = \begin{bmatrix} a_{-k-l+1} & a_{-k-l} & \cdots \\ a_{-k-l} & a_{-k-l-1} & \cdots \\ \cdot & \cdot & \cdots \end{bmatrix}$$

which is either Γ_{-k-l+2} or (if $k+l > 2$) a submatrix of Γ_0. In either case it is a contraction, and so T is itself a contraction, by Lemma 15.13. By Theorem 13.15 the function ψ with Fourier series

$$\psi(z) = \sum_{-\infty}^{\infty} a_n z^n$$

belongs to L^∞ and satisfies

$$\|\psi\|_\infty = \|T\| \leqslant 1.$$

Let

$$g = \varphi - \psi.$$

Since φ and ψ have the same negative Fourier coefficients, $g \in H^\infty$. Hence

$$\begin{aligned}
\operatorname{dist}_{L^\infty}(\varphi, H^\infty) &\leqslant \|\varphi - g\|_\infty \\
&= \|\psi\|_\infty \tag{15.4} \\
&\leqslant 1 \\
&= \|H_\varphi\|.
\end{aligned}$$

This proves (15.3). Further, since $g \in H^\infty$,

$$H_\psi = H_{\varphi - g} = H_\varphi$$

and

$$\|\psi\|_\infty = \|\varphi - g\|_\infty \leqslant \mathrm{dist}_{L^\infty}(\varphi, H^\infty).$$

(15.4) is the reverse inequality, and so

$$\|\psi\|_\infty = \mathrm{dist}_{L^\infty}(\varphi, H^\infty) = \|H_\varphi\|. \qquad \square$$

15.3   Solution of Nehari's problem

So if we are given $\varphi \in L^\infty$ and asked to approximate it as closely as possible in the L^∞-norm by an H^∞ function then the nearest we can get is $\|H_\varphi\|$. This does not conclude the solution of Nehari's problem, for we also require to find the best approximating H^∞ function g. The proof above suggests one method: calculate a_{-n}, $n \in \mathbb{N}$, generate $a_0, a_1, \ldots$ inductively by repeated applications of Parrott's theorem, and write

$$g(z) = \varphi(z) - \sum_{-\infty}^{\infty} a_n z^n.$$

This is hardly a practical procedure. For applications to engineering design a better way is needed. Luckily g can be produced more directly from an analysis of H_φ. If we were to mark scenic high points in the mode of Baedeker the result and proof which follow would surely deserve a couple of stars. That it is also useful makes it a particularly appealing piece of mathematics.

15.15 Definition   Let H and K be Hilbert spaces and let $T \in \mathcal{L}(H, K)$. A *maximizing vector for T* is a non-zero vector $x \in H$ such that

$$\|Tx\| = \|T\|\|x\|. \qquad \square$$

Thus a maximizing vector for T is one at which T attains its norm. On a Banach space, even rank 1 operators need not have maximizing vectors (Problem 6.6). The operator

$$(Mx)(t) = tx(t), \qquad 0 < t < 1,$$

is bounded on $L^2(0, 1)$ but has no maximizing vector. However, compact operators on Hilbert spaces do have maximizing vectors (Problem 15.8).

15.16 Theorem   Solution of Nehari's problem. Let $\varphi \in L^\infty$ and suppose that the Hankel operator

$$H_\varphi : H^2 \to L^2 \ominus H^2$$

has a maximizing vector f. There is a unique best approximation of $g \in H^\infty$ to φ in the L^∞-norm. It is given by

$$g = \varphi - \frac{H_\varphi f}{f}. \tag{15.5}$$

Proof. The statement is that there is a unique $g \in H^\infty$ such that

$$\|\varphi - g\|_\infty = \text{dist}_{L^\infty}(\varphi, H^\infty).$$

We may suppose $\|f\| = 1$. By Theorem 15.14 there exists $\psi \in L^\infty$ such that $H_\varphi = H_\psi$ and $\|\psi\|_\infty = \|H_\varphi\|$. We have

$$\begin{aligned}
\|H_\varphi\| = \|H_\varphi f\| &= \|H_\psi f\| \\
&= \|P_-(\psi f)\| \leqslant \|\psi f\| \\
&\leqslant \|\psi\|_\infty \|f\| = \|\psi\|_\infty \\
&= \|H_\varphi\|.
\end{aligned} \tag{15.6}$$

All the terms in the sequence must be equal: in particular,

$$\|P_-(\psi f)\| = \|\psi f\|.$$

Now

$$\psi f = P_-(\psi f) + P_+(\psi f),$$

P_+ being the orthogonal projection on H^2. The two terms on the right hand side are orthogonal, and so

$$\|\psi f\|^2 = \|P_-(\psi f)\|^2 + \|P_+(\psi f)\|^2.$$

Hence $P_+(\psi f) = 0$, and so

$$\psi f = P_-(\psi f) = H_\varphi f.$$

Since $f \in H^2$, the domain of H_φ, f is non-zero almost everywhere on $\partial \mathbb{D}$, by Theorem 13.13. We may therefore divide through by f to get

$$\psi(z) = \frac{(H_\varphi f)(z)}{f(z)} \quad \text{almost everywhere.} \tag{15.7}$$

Let

$$\begin{aligned}
g &= \varphi - \psi \\
&= \varphi - \frac{H_\varphi f}{f}.
\end{aligned}$$

Since $H_\varphi = H_\psi$, $g \in H^\infty$, and

$$\begin{aligned}
\|\varphi - g\|_\infty = \|\psi\|_\infty &= \|H_\varphi\| \\
&= \text{dist}_{L^\infty}(\varphi, H^\infty).
\end{aligned}$$

Thus (15.5) does give a best approximation $g \in H^\infty$ to φ in the L^∞-norm.

To prove uniqueness, let g be any best approximation: that is, $g \in H^\infty$ and $\|\varphi - g\|_\infty = \text{dist}_{L^\infty}(\varphi, H^\infty)$. Let $\psi = \varphi - g$. Then $H_\varphi = H_\psi$ and

$\|\psi\|_\infty = \|H_\varphi\|$, so that the above reasoning applies to show that ψ is given by (15.7).   □

Let us try calculating with this theorem.

15.17 Example Let $\alpha \in \mathbb{D}$ and let

$$\varphi(z) = \frac{1}{z-\alpha}, \qquad z \in \partial\mathbb{D}.$$

Find the best approximation $g \in H^\infty$ to φ with respect to the L^∞-norm.

The Fourier series of φ is

$$\varphi(z) = z^{-1} + \alpha z^{-2} + \alpha^2 z^{-3} + \cdots,$$

and so, with respect to the standard bases of H^2 and $L^2 \ominus H^2$ we have

$$H_\varphi \sim \begin{bmatrix} 1 & \alpha & \alpha^2 & \cdots \\ \alpha & \alpha^2 & \alpha^3 & \cdots \\ \cdot & \cdot & \cdot & \cdots \end{bmatrix}.$$

Thus H_φ has rank 1: in fact

$$H_\varphi x = (x, f)h$$

where $f \in H^2$, $h \in L^2 \ominus H^2$ have Fourier expansions

$$f \sim (1, \bar\alpha, \bar\alpha^2, \ldots), \qquad h \sim (1, \alpha, \alpha^2, \ldots).$$

Hence (cf. Problem 7.9)

$$\|H_\varphi\| = \|f\|\,\|h\| = 1 + |\alpha|^2 + |\alpha|^4 + \cdots$$
$$= (1 - |\alpha|^2)^{-1}.$$

Thus

$$\operatorname{dist}_{L^\infty}((z-\alpha)^{-1}, H^\infty) = (1 - |\alpha|^2)^{-1}.$$

A maximizing vector for H_φ is

$$f(z) = 1 + \bar\alpha z + \bar\alpha^2 z^2 + \cdots$$
$$= (1 - \bar\alpha z)^{-1},$$

and

$$H_\varphi f = (f, f)h = \frac{1}{1 - |\alpha|^2}\left(\frac{1}{z} + \frac{\alpha}{z^2} + \cdots\right)$$

$$= \frac{1}{1 - |\alpha|^2} \cdot \frac{1}{z - \alpha}.$$

The required best approximation g is given by

$$g(z) = \varphi(z) - \frac{(H_\varphi f)(z)}{f(z)}$$

$$= \frac{1}{z-\alpha} - \frac{1}{1 - |\alpha|^2} \frac{1 - \bar\alpha z}{z - \alpha}$$

$$= \frac{\bar\alpha}{1 - |\alpha|^2}.$$ □

In contrast, the best approximation in H^2 to φ with respect to the L^2-norm is the zero function, since $\varphi \perp H^2$.

The next simplest example to the above is in Problem 15.9. Anything more complicated than that gets too hard to calculate by hand since finding maximizing vectors is an eigenvalue problem. The method can be implemented on a computer fairly easily for rational functions φ. For non-rational φ there are severe numerical difficulties in representing the infinite-rank operator H_φ adequately and finding a maximizing vector. This is a topic of current research, having considerable practical significance.

It should be mentioned that the existence of a maximizing vector f is not guaranteed. There do exist functions $\varphi \in L^\infty$ having more than one best approximation in H^∞. In view of Theorem 15.16 (the uniqueness statement) the corresponding H_φ cannot have maximizing vectors. However, as long as φ is the sum of an H^∞ function and a continuous function on $\partial \mathbb{D}$, H_φ is compact and so does have maximizing vectors (Problems 15.5, 15.8). If φ is rational then H_φ has finite rank, and so finite matrix calculations suffice (Problem 15.9).

Theorem 15.16 was essentially proved by Z. Nehari in 1957. His statement was in terms of matrices rather than operators, and read as follows.

15.18 Nehari's theorem   The infinite Hankel matrix

$$H = [a_{-i-j+1}]_{i,j=1}^{\infty}$$

defines a bounded linear operator on ℓ^2 if and only if there exists $\varphi \in L^\infty$ such that $\hat{\varphi}(-n) = a_{-n}$, $n \in \mathbb{N}$.

When H is bounded its operator norm on ℓ^2 equals

$$\inf\{\|\varphi\|_\infty : \varphi \in L^\infty, \hat{\varphi}(-n) = a_{-n}, n \in \mathbb{N}\}.$$

Proof. It is tempting to argue that H is the matrix of H_φ, where

$$\varphi(z) = \sum_{-\infty}^{-1} a_n z^n, \tag{15.8}$$

and say that the result now follows from Theorem 15.14. However, there is a slight difficulty. Theorem 15.14, and indeed the definition of H_φ, only apply when $\varphi \in L^\infty$, and the example of Hilbert's Hankel matrix shows that the function (15.8) can be essentially unbounded. To prove the forward implication it is probably simplest to repeat the inductive construction using Parrott's theorem: filling in the details will make a good exercise for the reader. The reverse implication is easy. $\square$

15.4   Problems

15.1. Let M be a closed subspace of a Hilbert space H and let $P \in \mathcal{L}(H)$ be the orthogonal projection on H with range M (i.e. $Px = y$ if $x = y + z$ with $y \in M$, $z \in M^{\perp}$). Prove that P is Hermitian.

15.2. Let H be a Hilbert space and let $P \in \mathcal{L}(H)$ satisfy $P^2 = P = P^*$. Prove that P is an orthogonal projection on H with closed range.

15.3. Let $\varphi \in L^2$ be defined by

$$\varphi(e^{it}) = \begin{cases} 1 & \text{if } 0 < t < \pi, \\ -1 & \text{if } -\pi < t < 0. \end{cases}$$

Find the element of H^2 which is closest to φ in the L^2-norm.

15.4. Let p be a polynomial of degree $n \geqslant 1$ having all its zeros in $\mathbb{D}$, and let

$$\hat{p}(z) = z^n p(1/\bar{z})^{-}.$$

Show that if g is any polynomial of degree less than n then, for some polynomial $\tilde{g}$ and all $f \in L^2$,

$$\left(f, \frac{g}{p}\right) = \frac{1}{2\pi i} \int_{\partial \mathbb{D}} f(z) \frac{\tilde{g}(z)}{\hat{p}(z)}\, dz.$$

Deduce that $g/p \in L^2 \ominus H^2$.

Let $\varphi \in L^2$ be rational, and write

$$\varphi(z) = f(z) + \frac{g(z)}{p(z)} + \frac{h(z)}{q(z)}, \qquad z \in \partial \mathbb{D}$$

where f, g, h, p and q are polynomials, the degrees of g, h are less than the degrees of p, q respectively, the zeros of p lie in $\mathbb{D}$ and the zeros of q lie in the complement of clos $\mathbb{D}$. Show that the closest element of H^2 to φ (with respect to the L^2-norm) is $f + h/q$.

15.5. Show that if φ is a trigonometric polynomial (i.e. an element of the linear span of $\{z^n : n \in \mathbb{Z}\}$) then H_φ has finite rank. Hence show that if $\psi \in L^\infty$ is expressible in the form $\psi_1 + \psi_2$ with ψ_1 continuous on $\partial \mathbb{D}$ and $\psi_2 \in H^\infty$ then H_ψ is compact.

15.6. Let H, K be Hilbert spaces and let $T \in \mathcal{L}(H, K)$. Let x be a maximizing vector for T. Show that $\|T\|^2$ is an eigenvalue of T^*T and that x is a corresponding eigenvector. (Use the result of Problem 12.6.)

15.7. Let

$$\varphi(e^{i\theta}) = 1 + 2\cos\theta + 2\cos 2\theta, \qquad 0 < \theta < 2\pi.$$

Show that

$$\operatorname{dist}_{L^\infty}(\varphi, H^\infty) = \tfrac{1}{2}(1 + \sqrt{5})$$

and find the best approximation to φ in H^∞ with respect to the L^∞-norm.

15.8. Let H, K be Hilbert spaces and let $T \in \mathcal{L}(H, K)$, $T \neq 0$.

(i) Show that any maximizing vector for T^*T is also a maximizing vector for T.

(ii) Show that if x is a maximizing vector for T then Tx is a maximizing vector for T^*.

(iii) Show that if T is compact then T has a maximizing vector.

(iv) Show that if x is a maximizing vector for T then $x \perp \operatorname{Ker} T$ (use Lemma 4.23).

15.9. Let $\alpha \in \mathbb{D}$ and let

$$\varphi(z) = \frac{1}{z(z - \alpha)}, \qquad z \in \partial\mathbb{D}.$$

Find $\operatorname{dist}_{L^\infty}(\varphi, H^\infty)$, and find the best approximation to φ in H^∞ with respect to the L^∞-norm.

(Outline of a possible method. Let H be the Hankel matrix corresponding to the Hankel operator H_φ. Show that H has rank 2 and that its range has an orthonormal basis

$$e_1 = [1 \ \ 0 \ \ 0 \ \ \cdots]^\mathsf{T}, \qquad e_2 = \beta[0 \ \ 1 \ \ \alpha \ \ \alpha^2 \ \ \cdots]^\mathsf{T},$$

where $\beta = (1 - |\alpha|^2)^{1/2}$. With respect to the bases $(z^n)_{n=0}^\infty$ in H^2 and e_1, e_2 in range H, H_φ has matrix

$$K = \begin{bmatrix} 0 & 1 & \alpha & \cdots \\ 1/\beta & \alpha/\beta & \alpha^2/\beta & \cdots \end{bmatrix}.$$

It follows that $H_\varphi H_\varphi^*$ has matrix

$$KK^* = \frac{1}{\beta^4}\begin{bmatrix} \beta^2 & \bar\alpha\beta \\ \alpha\beta & 1 \end{bmatrix}$$

with respect to e_1, e_2. Hence $\|K\|$ can be found from the solution of a quadratic equation (recall Problems 7.21, 7.22). A maximizing vector for H_φ can be derived from an eigenvector for KK^* using Problem 15.8(ii).)

15.10. Let $\psi \in L^\infty$, $f \in L^2$ and suppose that

$$\|\psi f\|_{L^2} = \|\psi\|_\infty \|f\|_{L^2}.$$

Show that

$$|\psi(z)| = \|\psi\|_\infty \qquad \text{for almost all } z \in \partial\mathbb{D}\setminus f^{-1}(0).$$

15.11. Let $\varphi \in L^\infty$ and suppose that the Hankel operator H_φ has a maximizing vector. Let g be the unique best approximation to φ in H^∞ with respect to the L^∞-norm. Use relation (15.6) and the preceding problem to show that $|\varphi - g|$ is equal to a constant almost everywhere on $\partial\mathbb{D}$.

15.12. A model matching problem. Let

$$T_1(z) = 1 + 2z, \qquad T_2(z) = z(z - \tfrac{1}{2}), \qquad z \in \mathbb{D}.$$

Show that

$$\inf_{Q \in H^\infty} \| T_1 + T_2 Q \|_\infty = \operatorname{dist}_{L^\infty}(\varphi, H^\infty)$$

where

$$\varphi(z) = \frac{(1 + 2z)(1 - \tfrac{1}{2}z)}{z(z - \tfrac{1}{2})}$$

$$= -1 + \frac{z + 1}{z(z - \tfrac{1}{2})}.$$

Show that the vectors

$$e_1(z) = \frac{1}{z}, \qquad e_2(z) = \frac{\sqrt{3}}{2} \frac{1}{z(z - \tfrac{1}{2})}$$

constitute an orthonormal basis of Range H_φ, and that with respect to the bases $(z^n)_{n=0}^\infty$ of H^2, e_1, e_2 of Range H_φ, H_φ has matrix

$$\begin{bmatrix} 1 & 3/2 & 3/2^2 & 3/2^3 & \cdots \\ \sqrt{3} & \sqrt{3}/2 & \sqrt{3}/2^2 & \sqrt{3}/2^3 & \cdots \end{bmatrix}.$$

Deduce that

$$\inf_{Q \in H^\infty} \| T_1 + T_2 Q \|_\infty = 1 + \sqrt{3},$$

and find the unique function Q for which the infimum is attained.

15.13. **A Nevanlinna–Pick problem.** Show that the infimum of the H^∞-norms of all functions $f \in H^\infty$ which satisfy

$$f(0) = 1, \qquad f(\tfrac{1}{2}) = 2$$

is $1 + \sqrt{3}$, and find the unique interpolating function $f \in H^\infty$ of minimal norm.

Approximation by meromorphic functions

To conclude the book we shall prove a theorem which is in the same general area as the result on best analytic approximations in the last chapter but is more recent and harder. Once again we shall be seeking the best approximation (with respect to the essential supremum norm) of a given bounded function on the unit circle, but now, instead of requiring the approximating function to be analytic in $\mathbb{D}$, we shall allow it to have poles in $\mathbb{D}$, up to a prescribed number. This problem arises in engineering applications in the context of *model reduction*: that is, the problem of finding a simple model to replace a relatively complicated one without too great a loss of accuracy. The Adamyan–Arov–Krein theorem, which we present below, has had a considerable influence on the treatment of this problem. The theorem dates from 1968, but it is only since the end of the seventies that engineers realised they could use it (at least, in the west).

We shall say that a function $\psi \in L^\infty$ *has at most k poles in* $\mathbb{D}$ if there exist $g \in H^\infty$, $r \leqslant k$ and $\alpha_1, \ldots, \alpha_r \in \mathbb{D}$ such that

$$\psi(z) = \frac{g(z)}{(z - \alpha_1) \cdots (z - \alpha_r)} \qquad \text{almost everywhere on } \partial\mathbb{D}.$$

We denote by $H^\infty_{(k)}$ the set of functions in L^∞ having at most k poles in $\mathbb{D}$. The problem we study is:

Given $\varphi \in L^\infty$ *and a non-negative integer k, find*

$$\inf\{\|\varphi - \psi\|_\infty : \psi \in H^\infty_{(k)}\}$$

and a function $\psi \in H^\infty_{(k)}$ *for which the infimum is attained.*

We can write the infimum as $\mathrm{dist}_{L^\infty}(\varphi, H^\infty_{(k)})$. In the case $k = 0$ the problem reduces to the one solved in the last chapter. For the more general problem we need analogues of the notions of operator norm and maximizing vector.

16.1   The singular values of an operator

16.1   *Definition*   Let H, K be Hilbert spaces and let $T \in \mathcal{L}(H, K)$. For any non-negative integer k let

$$s_k(T) = \inf\{\|T - R\| : R \in \mathcal{L}(H, K), \text{rank } R \leqslant k\}.$$

The numbers

$$s_0(T) \geqslant s_1(T) \geqslant s_2(T) \geqslant \cdots \geqslant 0$$

are called the *s-numbers* or *singular values* of *T*.   □

Thus $s_k(T)$ is the distance, with respect to the operator norm, of T from the set of operators of rank at most k in $\mathcal{L}(H, K)$. Here is a more 'spatial' characterization.

16.2 *Theorem*   Let H, K and T be as in Definition 16.1. Then

$$s_k(T) = \inf_E \|T \mid E\|,$$

where $T \mid E$ denotes the restriction of T to E, and E ranges over the closed subspaces of H of codimension at most k.

Proof. Let E be a closed subspace of H of codimension at most k. Define an operator $R \in \mathcal{L}(H, K)$ by

$$Rx = \begin{cases} Tx & \text{if } x \in E^{\perp}, \\ 0 & \text{if } x \in E. \end{cases}$$

Since $\dim E^{\perp} \leqslant k$, rank $R \leqslant k$. Since $T - R$ is zero on $E^{\perp}$ we have

$$\|T - R\| = \|T - R \mid E\|$$
$$= \|T \mid E\|.$$

Hence

$$\|T \mid E\| \geqslant s_k(T),$$

and so

$$\inf_{\text{codim } E \leqslant k} \|T \mid E\| \geqslant s_k(T).$$

Now pick any $\varepsilon > 0$ and an operator R of rank at most k such that

$$\|T - R\| < s_k(T) + \varepsilon.$$

The restriction of R to $(\text{Ker } R)^{\perp}$ is a linear bijection between $(\text{Ker } R)^{\perp}$ and Range R, from which it follows that Ker R has codimension at most k in H. And

$$\|T \mid \text{Ker } R\| = \|(T - R) \mid \text{Ker } R\|$$
$$\leqslant \|T - R\|$$
$$< s_k(T) + \varepsilon.$$

Hence

$$\inf_{\text{codim } E \leqslant k} \|T \mid E\| \leqslant s_k(T).$$ □

16.3 Corollary   Suppose that

$$\|Tx\| \geqslant t\|x\|$$

for all $x \in F$, where F is a subspace of H of dimension $n + 1$, n being a non-negative integer. Then $s_n(T) \geqslant t$.

Proof. Let E be any closed subspace of H of codimension n, so that the quotient space H/E has dimension n. If $E \cap F = \{0\}$ then the restriction to F of the quotient mapping $H \to H/E$ is injective, and hence maps the $(n + 1)$-dimensional space F injectively into the n-dimensional space H/E. This is impossible, and so $E \cap F$ contains a non-zero vector x. Since $\|Tx\|/\|x\| \geqslant t$, we have $\|T \mid E\| \geqslant t$. Hence

$$s_n(T) \geqslant t. \qquad \qquad \qquad \square$$

To avoid some minor technicalities we restrict attention to compact T.

16.4 Theorem   Let T be a compact operator from H to K, where H, K are Hilbert spaces. Let $(e_n)_0^\infty$ be a complete orthonormal sequence in H consisting of eigenvectors of T^*T, numbered so that the corresponding eigenvalues are in non-increasing order. Then the eigenvalue of T^*T corresponding to e_n is $s_n(T)^2$.

Proof. Let E be the closed linear span of $\{e_j : j \geqslant k\}$. Then $E^\perp$ is spanned by $e_0, \ldots, e_{k-1}$ and so is of dimension k. We have

$$\|T \mid E\|^2 = \sup\{(T^*Tx, x) : x \in E, \|x\| \leqslant 1\}.$$

If λ_n denotes the eigenvalue of the compact Hermitian operator T^*T corresponding to the eigenvector e_n, then the right hand side is clearly λ_k. Thus, by Theorem 16.2,

$$s_k(T) \leqslant \|T \mid E\| = \lambda_k^{1/2}.$$

Now consider any closed subspace E of codimension at most k in H. The space

$$F = \text{lin}\{e_0, e_1, \ldots, e_k\}$$

has dimension $k + 1$, and so has non-trivial intersection with E. Let x be a non-zero vector in $E \cap F$: say

$$x = \sum_{j=0}^k x_j e_j.$$

Then

$$\|Tx\|^2 = (T^*Tx, x) = \sum_0^k \lambda_j |x_j|^2 \geqslant \lambda_k \|x\|^2.$$

Hence $\|T \mid E\| \geqslant \lambda_k^{1/2}$ for any E of codimension at most k in H. Thus

$$s_k(T) = \lambda_k^{1/2}, \qquad k = 0, 1, 2, \ldots . \qquad \qquad \square$$

This result, together with the spectral theorem, shows that the singular values of T are the square roots of the eigenvalues of T^*T as long as H is separable and infinite-dimensional. If H is not separable we may apply the result to $(\mathrm{Ker}\, T)^{\perp}$, which is, to obtain the same conclusion.

16.2   Schmidt pairs and singular vectors

Suppose s is a singular value of a compact operator $T \in \mathcal{L}(H, K)$. Then s^2 is an eigenvalue of T^*T and so there is a corresponding eigenvector $x \in H$:

$$T^*Tx = s^2x.$$

If $s \neq 0$ we may let $y = s^{-1}Tx \in K$, and then $T^*y = s^{-1}T^*Tx = sx$.

16.5 Definitions   Let H, K be Hilbert spaces and let $T \in \mathcal{L}(H, K)$. Let s be a singular value of T. A *Schmidt pair* for T corresponding to s is a pair $\langle x, y \rangle$ of non-zero vectors, with $x \in H$, $y \in K$, such that

$$Tx = sy, \qquad T^*y = sx.$$

A *singular vector* or *Schmidt vector* for T corresponding to s is an eigenvector of T^*T corresponding to s^2. The *multiplicity* of a singular value s of T is the dimension of the eigenspace of T^*T corresponding to s^2, i.e. $\dim \mathrm{Ker}(s^2I - T^*T)$. $\square$

If s is a non-zero singular value of T of multiplicity 1 then clearly the corresponding Schmidt pair is uniquely determined up to a non-zero scalar multiple: if s has multiplicity m then the space of Schmidt vectors has dimension m.

Suppose T is compact and $(e_n)_0^{\infty}$ is a complete orthonormal sequence in $(\mathrm{Ker}\, T)^{\perp}$ consisting of eigenvectors of T^*T, so that

$$T^*Te_n = s_n^2e_n, \qquad s_n > 0.$$

Let $f_n = s_n^{-1}Te_n$, so that $\langle e_n, f_n \rangle$ is a Schmidt pair for T corresponding to s_n. We can write an arbitrary element of H as $x = \sum_0^{\infty} x_ne_n + x'$ with $x_n \in \mathbb{C}$, $x' \in \mathrm{Ker}\, T$. Then the action of T can be described by

$$Tx = \sum_0^{\infty} s_nx_nf_n. \tag{16.1}$$

This representation of a non-Hermitian compact operator T can be as useful for some purposes as that given by the spectral theorem for Hermitian ones; however, for many purposes it is not so easily applied (for example, if $H = K$ and we wanted to find a non-Hermitian square root of T: (16.1) would not help). The reader who is familiar with singular value decompositions of matrices should work out how the two notions correspond (Problem 16.1).

16.6 Lemma   Let H, K be separable Hilbert spaces and let T be a compact operator from H to K. Let $k \in \mathbb{Z}^+$ and let E be a closed subspace of H of codimension at most k such that

$$\|T \mid E\| = s_k(T).$$

Then E contains a singular vector of T corresponding to $s_k(T)$.

Proof. Let (e_n) be a complete orthonormal sequence in H consisting of eigenvectors of T^*T, numbered so that the corresponding eigenvalues s_n^2, where $s_n \geqslant 0$, are in non-increasing order. Suppose that

$$s_0 \geqslant s_1 \geqslant \cdots \geqslant s_{l-1} > s_l = \cdots = s_k \geqslant \cdots$$

where $0 \leqslant l \leqslant k$. E has codimension $\leqslant k$ and so has non-trivial intersection with the linear span of the $k + 1$ orthonormal vectors $e_0, \ldots, e_k$. Pick a unit vector

$$x = \lambda_0 e_0 + \cdots + \lambda_k e_k \in E.$$

Since $\|T \mid E\| = s_k$ we have

$$s_k^2 \sum_{j=0}^{k} |\lambda_j|^2 = s_k^2$$

$$\geqslant \|Tx\|^2 = (T^*Tx, x)$$

$$= \sum_{j=0}^{k} s_j^2 |\lambda_j|^2$$

$$\geqslant s_{l-1}^2 \sum_{j<l} |\lambda_j|^2 + s_k^2 \sum_{j=l}^{k} |\lambda_j|^2.$$

Hence

$$(s_{l-1}^2 - s_k^2) \sum_{j<l} |\lambda_j|^2 \leqslant 0,$$

and so we must have $\lambda_0 = \cdots = \lambda_{l-1} = 0$. Hence x belongs to $\mathrm{lin}\{e_l, \ldots, e_k\}$, and so is a singular vector of T corresponding to $s_k = s_k(T)$. $\square$

16.7 Lemma   Let x be a singular vector of an operator T corresponding to a non-zero singular value. Then $x \in (\mathrm{Ker}\ T)^\perp$.

Proof. If $y \in \mathrm{Ker}\ T$ then x, y are eigenvectors of the Hermitian operator T^*T with respect to distinct eigenvalues, and hence $x \perp y$. $\square$

We shall need some generalities about rational functions and Hankel operators. Recall from Problem 13.5 that a function of the form

$$B(z) = \lambda \prod_{j=1}^{n} \frac{z - \alpha_j}{1 - \bar{\alpha}_j z},$$

where $n \in \mathbb{Z}^+, \alpha_1, \ldots, \alpha_n \in \mathbb{D}$ and λ is a complex number of unit modulus, is called a *finite Blaschke product* (in the case $n = 0$ the product is empty, and is conventionally interpreted as 1, so that the constant function $B \equiv \lambda$ is also regarded as a Blaschke product). Here n is called the *degree* of B. It is easy to see that B is an inner function.

16.8 Exercise   Let f be a rational inner function. Show that f is a finite Blaschke product.   □

If τ, υ are finite Blaschke products we say that τ *divides* υ if there exists $\sigma \in H^\infty$ such that $\upsilon = \tau\sigma$. Such a σ must itself be a finite Blaschke product.

Consider an arbitrary rational function

$$\varphi(z) = \frac{c(z - \beta_1)(z - \beta_2) \dots (z - \beta_r)}{(z - \alpha_1)(z - \alpha_2) \dots (z - \alpha_m)}$$

where $c \in \mathbb{C}$ and the expression is in its lowest terms, so that no β coincides with an α. If the restriction of φ to $\partial\mathbb{D}$ is an L^2 function then φ can have no poles on $\partial\mathbb{D}$, and so the αs split up into n points in $\mathbb{D}$ and $m - n$ in $\mathbb{C}\backslash\text{clos }\mathbb{D}$. We may suppose them numbered so that $\alpha_1, \dots, \alpha_n \in \mathbb{D}$. We can then write

$$\varphi(z) = \frac{1}{(z - \alpha_1) \dots (z - \alpha_n)} \cdot \frac{c(z - \beta_1) \dots (z - \beta_r)}{(z - \alpha_{n+1}) \dots (z - \alpha_m)}$$

$$= \frac{(1 - \bar{\alpha}_1 z) \dots (1 - \bar{\alpha}_n z)}{(z - \alpha_1) \dots (z - \alpha_n)} \cdot \frac{c(z - \beta_1) \dots (z - \beta_r)}{(1 - \bar{\alpha}_1 z) \dots (1 - \bar{\alpha}_m z)}.$$

In both of these expressions for φ the second factor is a rational H^∞ function. The following definitions therefore make sense.

16.9 Definition   Let φ be a rational function whose restriction to $\partial\mathbb{D}$ is an L^2 function. The *stable denominator* of φ is the unique monic polynomial p such that

(i) all zeros of p lie in $\mathbb{D}$;

(ii) φ is expressible in the form g/p where $g \in H^\infty$ and p, g have no common zero.

The *inner denominator* of φ is the finite Blaschke product

$$B = p/\hat{p}$$

where $\hat{p}$ is the polynomial

$$\hat{p}(z) = z^n p(1/\bar{z})^-, \qquad z \in \mathbb{C}\backslash\{0\}.$$ □

Alternatively, the inner denominator B of φ is characterized by the properties (i) B is a finite Blaschke product, (ii) $B\varphi \in H^\infty$, (iii) B, φ have no common zero. In particular, if $\varphi \in H^\infty$, $p = B = 1$.

This terminology is not standard, but polynomials having all their zeros in $\mathbb{D}$ are sometimes called *stable polynomials*.

16.10 Exercise   Let φ be a rational function with no poles on $\partial\mathbb{D}$. Prove that $\varphi \in H^{\infty}_{(k)}$ if and only if the inner denominator of φ has degree at most k. ☐

If $g \in H^{\infty}$ we write gH^2 for the subspace $\{gx : x \in H^2\}$ of H^2.

16.11 *Lemma* Let φ be a rational function having no poles on $\partial\mathbb{D}$. Then

$$\text{Ker } H_{\varphi} = pH^2 = vH^2$$

where p, v are the stable denominator and inner denominator of φ respectively.

Proof. Let $x \in H^2$. Then $x \in \text{Ker } H_{\varphi}$ if and only if $P_-(\varphi x) = 0$, i.e. $\varphi x \in H^2$. Write $\varphi = g/p$ with g, p as in Definition 16.9. If $x \in pH^2$ then $\varphi x = gx/p \in H^2$. Conversely, suppose $\varphi x \in pH^2$. Write $g = g_1/g_2$ where g_1, g_2 are relatively prime polynomials. Then

$$\frac{g_1 x}{g_2 p} = \varphi x \in H^2,$$

and so $g_1 x = ph$ for some $h \in H^2$. Since g, p have no common zero, g_1 and p are relatively prime polynomials, so that there exist polynomials a, b such that

$$ag_1 + bp = 1.$$

Then

$$x = ag_1 x + bpx = aph + bpx$$
$$= p(ah + bx) \in pH^2.$$

Hence $\text{Ker } H_{\varphi} = pH^2$. Since both $\hat{p}$ and $1/\hat{p}$ are H^{∞} functions, $H^2 = (1/\hat{p})H^2$ and so $pH^2 = vH^2$. ☐

As in the last chapter we denote by P_+ the orthogonal projection operator from L^2 to H^2.

16.12 *Exercise* Show that $H^*_{\varphi} : L^2 \ominus H^2 \to H^2$ is given by

$$H^*_{\varphi} x = P_+(\bar{\varphi} x), \qquad x \in L^2 \ominus H^2. \qquad ☐$$

There is a notational point relating to complex conjugation which deserves elucidation. We saw in Chapter 13 that H^{∞}, defined as a space of (equivalence classes of) functions on $\partial\mathbb{D}$, could be identified with the space of all bounded analytic functions in $\mathbb{D}$, and we have used a single symbol (f, say) to denote either type of function at will. We also use the symbol $\tilde{f}$, defined by $\tilde{f}(z) = \overline{f(z)}$. There is an ambiguity here. Consider the question: if f is rational, does it follow that $\tilde{f}$ is rational? If we are thinking of functions on $\partial\mathbb{D}$, the answer is yes since $z\bar{z} = 1$. For example, if $f(z) = z^3/(z^2 + 2)$ then

$$\tilde{f}(z) = \frac{\bar{z}^3}{\bar{z}^2 + 2} = \frac{1}{z + 2z^3}, \qquad z \in \partial\mathbb{D}.$$

However, if we are speaking of functions on $\mathbb{D}$ the answer is no: if $f(z) = z$ then $\bar{f}$ is not a rational function on $\mathbb{D}$. Here we shall use the notation $\bar{f}$ for the function f defined (almost everywhere) on $\partial \mathbb{D}$. Thus, if v is an inner function we have $v\bar{v} = 1$. This is clearly true on $\partial \mathbb{D}$, but false if interpreted as meaning $v(z)\bar{v}(z) = 1$ for all $z \in \mathbb{D}$. I shall further presume on the goodwill of the reader by writing zg, $\bar{z}g$ to denote the functions on $\partial \mathbb{D}$ whose values at the point z are $zg(z)$, $\bar{z}g(z)$ respectively.

16.13 *Kronecker's theorem* Let $\varphi \in L^\infty$. Then H_φ has finite rank $\leqslant k$ if and only if $\varphi \in H^\infty_{(k)}$.

Proof. Suppose rank $H_\varphi \leqslant k$. The $k + 1$ vectors $H_\varphi 1, H_\varphi z, \ldots, H_\varphi z^k$ in Range H_φ are linearly dependent, and hence there exists a non-zero polynomial p of degree at most k such that $H_\varphi p = 0$. That is, $\varphi p \in H^2$. Since both φ and p are essentially bounded, $\varphi p \in H^\infty$. Thus φ is expressible as g/p with $g \in H^\infty$ and p of degree $\leqslant k$, and so $\varphi \in H^\infty_{(k)}$.

Conversely, suppose $\varphi \in H^\infty_{(k)}$, so that $p\varphi \in H^\infty$ for some polynomial p of degree $\leqslant k$ having all its zeros in $\mathbb{D}$. Then $\varphi x \in H^2$ for any $x \in pH^2$: that is, $pH^2 \subset \text{Ker } H_\varphi$. Thus $(\text{Ker } H_\varphi)^\perp$ is contained in $H^2 \ominus pH^2$, which (by Problem 13.12) has dimension at most k. Now for any operator T between Hilbert spaces, the restriction of T to $(\text{Ker } T)^\perp$ is bijective onto range T, so that rank $T = \dim(\text{Ker } T)^\perp$. It follows that rank $H_\varphi \leqslant k$. $\square$

This brings us to the main result of the chapter.

16.3   The Adamyan–Arov–Krein theorem

16.14   *Theorem of Adamyan, Arov and Krein* Let $\varphi \in L^\infty$ be a rational function and let k be a non-negative integer. Then

$$\text{dist}_{L^\infty}(\varphi, H^\infty_{(k)}) = s_k(H_\varphi). \tag{16.2}$$

Furthermore this distance is attained at a unique function $\psi \in H^\infty_{(k)}$, and if f is a singular vector of H_φ corresponding to $s_k(H_\varphi)$ then

$$\psi = \varphi - \frac{H_\varphi f}{f}. \tag{16.3}$$

Proof. We may assume $k > 0$ since the case $k = 0$ is precisely Theorems 15.14 and 15.16. A natural idea is to try to reduce the general case to this known case, and the use of inner denominators will allow us to pass from $H^\infty_{(k)}$ functions to H^∞ functions without changing L^∞-norms.

To prove the easier inequality $(\geqslant)$ consider an arbitrary $g \in H^\infty_{(k)}$. We can write

$$g(z) = \frac{G(z)}{(z - \alpha_1) \ldots (z - \alpha_r)}$$

where $r \leqslant k$, $\alpha_1, \ldots, \alpha_r \in D$ and $G \in H^\infty$. Let

$$B(z) = \prod_{j=1}^{r} \frac{z - \alpha_j}{1 - \bar{\alpha}_j z}.$$

Then $Bg \in H^\infty$ and $|B(z)| = 1$ on ∂D. Hence

$$
\begin{aligned}
\|\varphi - g\|_\infty &= \|B(\varphi - g)\|_\infty \\
&= \|B\varphi - Bg\|_\infty \\
&\geqslant \mathrm{dist}_{L^\infty}(B\varphi, H^\infty) \\
&= \|H_{B\varphi}\|,
\end{aligned}
\tag{16.4}
$$

by Theorem 15.14. Now

$$
\begin{aligned}
\|H_{B\varphi}\| &= \sup_{|x| \leqslant 1} \|P_-(B\varphi x)\| \\
&= \sup_{|y| \leqslant 1, y \in BH^2} \|P_-(\varphi y)\| \\
&= \|H_\varphi \,|\, BH^2\|.
\end{aligned}
\tag{16.5}
$$

Since BH^2 has codimension $r \leqslant k$ in H^2 (see Problem 13.12 or 16.7), Theorem 16.2 yields

$$\|H_{B\varphi}\| = \|H_\varphi \,|\, BH^2\| \geqslant s_k(H_\varphi). \tag{16.6}$$

Thus

$$\|\varphi - g\|_\infty \geqslant s_k(H_\varphi).$$

Taking the infimum over $g \in H^\infty_{(k)}$ we have

$$\mathrm{dist}_{L^\infty}(\varphi, H^\infty_{(k)}) \geqslant s_k(H_\varphi).$$

To prove the opposite inequality we need to construct $\psi \in H^\infty_{(k)}$ such that $\|\varphi - \psi\|_\infty = s_k(H_\varphi)$. The statement of the theorem tells us which function to try, but it is unsatisfactory to pull formula (16.3) out of thin air. Let us see how we might arrive at (16.3) using Theorem 15.16.

Suppose, then, that we have guessed the distance formula (16.2) and chosen a function $\psi \in H^\infty_{(k)}$ such that

$$\|\varphi - \psi\|_\infty = s_k(H_\varphi). \tag{16.7}$$

As in the first part of the proof we take B to be the inner denominator of ψ, so that $B\psi \in H^\infty$. For brevity let us write s for $s_k(H_\varphi)$. I claim that $B\psi$ is a best approximation in H^∞ to $B\varphi$. Putting $g = \psi$ in (16.4) and (16.7) yields

$$
\begin{aligned}
\|\varphi - \psi\|_\infty &= \|B\varphi - B\psi\|_\infty \geqslant \|H_{B\varphi}\| \\
&= \mathrm{dist}_{L^\infty}(B\varphi, H^\infty) \geqslant s = \|\varphi - \psi\|_\infty.
\end{aligned}
\tag{16.8}
$$

Hence

$$\|B\varphi - B\psi\|_\infty = \mathrm{dist}_{L^\infty}(B\varphi, H^\infty),$$

as desired. Since $B\varphi$ is rational, $H_{B\varphi}$ has finite rank, by Kronecker's theorem. It therefore has a maximizing vector $x \in H^2$ (cf. Problem 15.8(iii)).

Theorem 15.16 now tells us that $B\psi$ is uniquely determined and is given by

$$B\psi = B\varphi - \frac{H_{B\varphi}x}{x}. \qquad (16.9)$$

Divide by B to get

$$\psi = \varphi - \frac{H_{B\varphi}x}{Bx}.$$

Since

$$H_{B\varphi}x = P_-(\varphi Bx) = H_\varphi(Bx),$$

we obtain

$$\psi = \varphi - \frac{H_\varphi(Bx)}{Bx}. \qquad (16.10)$$

Now (16.8) shows that

$$\text{dist}_{L^\infty}(B\varphi, H^\infty) = s,$$

and so, by (16.5),

$$\|H_\varphi \mid BH^2\| = \|H_{B\varphi}\|$$
$$= \text{dist}_{L^\infty}(B\varphi, H^\infty) = s.$$

BH^2 has codimension $\leqslant k$ in H^2, and so, by Lemma 16.6, BH^2 contains a singular vector f for H_φ corresponding to s. Write $f = B\xi$, $\xi \in H^2$. Then

$$\|H_{B\varphi}\xi\| = \|H_\varphi(B\xi)\| = \|H_\varphi f\|$$
$$= s\|f\| = s\|\xi\|.$$

Thus ξ is a maximizing vector for $H_{B\varphi}$. Now Theorem 15.16 tells us that in the formula (16.9) for $B\psi$ we may take x to be *any* maximizing vector of $H_{B\varphi}$, so we may assume x was chosen to be ξ. Then $f = Bx$ and so (16.10) becomes

$$\psi = \varphi - \frac{H_\varphi f}{f},$$

valid whenever f is a singular vector of H_φ corresponding to s and $f \in BH^2$.

   To summarize, on the assumption that the distance formula (16.2) is correct and that the distance is attained we have shown that any best approximation ψ in $H_{(k)}^\infty$ to φ is indeed given by (16.3), for *some* singular vector f of H_φ corresponding to s. The theorem states that f may be any such vector: we shall have to show they all give rise to the same function ψ.

   Before we resume the formal proof of Theorem 16.14 let us note another consequence that will follow if the distance formula is true. By Problem 15.11, $B\psi - B\varphi$ has constant modulus s on $\partial \mathbb{D}$ (almost everywhere), and the same therefore holds for $\psi - \varphi$: that is,

$$|(H_\varphi f)(z)| = s|f(z)|, \qquad z \in \partial \mathbb{D}.$$

In fact we need to establish this relation as a step towards proving the distance formula.

16.15 Lemma Let $\varphi \in L^\infty$ and suppose f is a singular vector of H_φ corresponding to a singular value s. Then

$$|(H_\varphi f)(z)| = s|f(z)| \tag{16.11}$$

for almost all $z \in \partial D$.

Proof. It suffices to show that the integrable functions $|H_\varphi f(\cdot)|^2$ and $s^2|f(\cdot)|^2$ have the same Fourier coefficients. For any $g \in L^2$ the nth Fourier coefficient of $|g(\cdot)|^2$ is given by

$$\frac{1}{2\pi} \int_{-\pi}^{\pi} |g(e^{i\theta})|^2 \, e^{-in\theta} \, d\theta = (g, z^n g)_{L^2}.$$

Thus, if c_n denotes the nth Fourier coefficient of $s^2|f(\cdot)|^2$ we have

$$c_n = (sf, z^n sf) = (s^2 f, z^n f).$$

Since f is a singular vector of H_φ corresponding to s,

$$H_\varphi^* H_\varphi f = s^2 f.$$

Thus

$$c_n = (H_\varphi^* H_\varphi f, z^n f) = (H_\varphi f, H_\varphi (z^n f))$$
$$= (H_\varphi f, P_-(z^n \varphi f)).$$

Since $H_\varphi f \perp H^2$ we have

$$(H_\varphi f, P_- g) = (H_\varphi f, g)$$

for any $g \in L^2$, and so

$$c_n = (H_\varphi f, z^n \varphi f)$$
$$= (\bar{z}^n H_\varphi f, \varphi f).$$

If now $n \geqslant 0$ we have $\bar{z}^n H_\varphi f \perp H^2$ and so

$$c_n = (\bar{z}^n H_\varphi f, P_-(\varphi f))$$
$$= (\bar{z}^n H_\varphi f, H_\varphi f)$$
$$= (H_\varphi f, z^n H_\varphi f), \tag{16.12}$$

which is the nth Fourier coefficient of $|(H_\varphi f)(\cdot)|^2$. Thus the nth Fourier coefficients of $|(H_\varphi f)(\cdot)|^2$ and $s^2|f(\cdot)|^2$ coincide when $n \geqslant 0$. However, for any real-valued integrable function F on ∂D,

$$\hat{F}(-n) = \hat{F}(n)^-.$$

Hence the negative Fourier coefficients also coincide. The functions are therefore equal almost everywhere on ∂D (see Rudin, 1966, Theorem 5.15), and (16.11) is established. $\qquad\square$

Now let us embark on the proof of the harder inequality ($\leqslant$) in the distance formula (16.2). By Kronecker's theorem H_φ has finite rank, and so there exists a singular vector f of H_φ corresponding to the singular value s. Let

$$\psi = \varphi - (H_\varphi f)/f.$$

By Theorem 13.13 the second term on the right hand side exists almost everywhere on $\partial\mathbb{D}$, and by Lemma 16.15 its modulus is s almost everywhere. Thus $\psi \in L^\infty$ and $\|\psi - \varphi\|_\infty = s$. The distance formula will follow if we can show $\psi \in H^\infty_{(k)}$: this is the hardest part of the proof.

Let B be the inner denominator of ψ and let n be the degree of B. We must show that $n \leqslant k$. To see the plan of campaign, suppose for the moment that s is a singular value of H_φ of multiplicity 1, so that

$$\cdots \geqslant s_{k-1}(H_\varphi) > s > s_{k+1}(H_\varphi) \geqslant \cdots.$$

If we can find a subspace F of H^2 of dimension $n + 1$ such that

$$\|H_\varphi x\| \geqslant s\|x\|, \qquad \text{all } x \in F, \tag{16.13}$$

then it will follow from Corollary 16.3 that $s_n(H_\varphi) \geqslant s_k$, and this can only happen if $n \leqslant k$. Much the same idea applies when s has multiplicity $m \geqslant 1$: say

$$\cdots s_{l-1} > s_l = \cdots = s_{k-1} = s = s_{k+1} = \cdots = s_{l+m-1} > s_{l+m} \geqslant \cdots.$$

Now we must find $F \subset H^2$ of dimension $n + m$ satisfying (16.13). Then we shall have $s_{n+m-1} \geqslant s$, and so

$$n + m - 1 \leqslant l + m - 1, \tag{16.14}$$

giving $n \leqslant l \leqslant k$.

We took B to be the inner denominator of ψ: we had better check that ψ is rational. If $s = 0$ then $H_\varphi f = 0$ and so $\psi = \varphi$, which is rational by hypothesis. Otherwise, by Lemma 16.7, $f \perp \operatorname{Ker} H_\varphi$; that is, by Lemma 16.11, $f \in H^2 \ominus qH^2$ where q is the stable denominator of φ. By Problem 13.12, f is a rational function. Then

$$H_\varphi f = P_-(\varphi f),$$

and, as we saw in Problem 15.4 P_- maps rational functions to rational functions. Thus $H_\varphi f$ is also rational. It follows that $\psi = \varphi - H_\varphi f/f$ is rational.

We shall construct $F \subset H^2$ of dimension $n + m$ such that

$$\|H_{B\varphi} x\| = s\|x\|, \qquad \text{all } x \in F. \tag{16.15}$$

Now for any $y \in L^2$ we have

$$\|P_-(By)\| \leqslant \|P_- y\|.$$

For if we write $y = y_+ + y_-$ with $y_+ \in H^2$, $y_- \perp H^2$ then $By_+ \in H^2$ and so

$$\|P_-(By)\| = \|P_-(By_+) + P_-(By_-)\|$$
$$= \|P_-(By_-)\| \leqslant \|By_-\|$$
$$= \|y_-\| = \|P_- y\|.$$

Hence, for $x \in H^2$,

$$\|H_{B\varphi} x\| = \|P_-(B\varphi x)\|$$
$$\leqslant \|P_-(\varphi x)\| = \|H_\varphi x\|.$$

Thus, if F satisfies (16.15) we have, for $x \in F$,

$$\|H_\varphi x\| \geqslant \|H_{B\varphi} x\| = s\|x\|,$$

and it follows by the foregoing argument that $n \leqslant k$, and so $\psi \in H_{(k)}^\infty$.

It remains to exhibit an $(n + m)$-dimensional subspace F of H^2 satisfying (16.15). We shall produce $n + m$ linearly independent singular vectors for $H_{B\varphi}$ corresponding to s: their linear span will be a suitable F.

If $s = 0$ then H_φ has rank at most k. By Kronecker's theorem $\varphi \in H_{(k)}^\infty$, so that both sides are zero in the distance formula, and the unique best approximation to φ in $H_{(k)}^\infty$ is given by $\psi = \varphi$, verifying (16.3). We may therefore assume that $s \neq 0$. Let $g = s^{-1} H_\varphi f$, so that

$$H_\varphi f = sg, \qquad H_\varphi^* g = sf.$$

That is, $f \in H^2$, $g \in L^2 \ominus H^2$ and

$$P_-(\varphi f) = sg, \qquad P_+(\bar\varphi g) = sf,$$

or equivalently

$$\varphi f = x + sg, \qquad \bar\varphi g = sf + y \tag{16.16}$$

for some $x \in H^2$, $y \in L^2 \ominus H^2$. By choice of ψ,

$$\varphi = \psi + sg/f,$$
$$\bar\varphi = \psi^- + s(g/f)^- = \psi^- + s(f/g),$$

the latter step following from the fact that g/f is unimodular by (16.11). Hence

$$\varphi f = \psi f + sg, \qquad \bar\varphi g = \psi^- g + sf. \tag{16.17}$$

Comparing these equations with (16.16) we find,

$$\psi f = x \in H^2, \qquad \psi^- g = y \in L^2 \ominus H^2.$$

The first of these gives us $f \in \operatorname{Ker} H_\psi$, and so by Lemma 16.11, $f \in BH^2$. The second gives $\psi \bar g \in zH^2$, hence $\bar z g \in \operatorname{Ker} H_\psi = BH^2$, and so

$$f \in BH^2, \qquad g \in \bar B(L^2 \ominus H^2). \tag{16.18}$$

Next we show that if C is an inner divisor of B then $\bar C f$ is a singular vector of $H_{B\varphi}$ corresponding to s. Write $B = CD$, where C, D are finite Blaschke products. I claim that $\langle \bar{C}f, Dg \rangle$ is a Schmidt pair of $H_{B\varphi}$ corresponding to s. It follows from (16.18) that $\bar{C}f \in H^2$, $Dg \in L^2 \ominus H^2$. Moreover

$$H_{B\varphi}\bar{C}f = P_-(\varphi B\bar{C}f)$$
$$= P_-((\psi + sg/f)B\bar{C}f)$$
$$= P_-(B\psi\bar{C}f) + P_-(sDg).$$

Since $B\psi \in H^\infty$, $\bar{C}f \in H^2$ and $Dg \perp H^2$ we obtain

$$H_{B\varphi}\bar{C}f = sDg. \tag{16.19}$$

Similarly,

$$H_{B\varphi}^*(Dg) = P_+(\bar{\varphi}\bar{B}Dg)$$
$$= P_+((\psi^- + sf/g)\bar{B}Dg)$$
$$= P_+((B\psi)^- Dg) + P_+(s\bar{C}f)$$
$$= P_+(s\bar{C}f) = s\bar{C}f,$$

that is,

$$\langle \bar{C}f, Dg \rangle \text{ is a Schmidt pair of } H_{B\varphi}. \tag{16.20}$$

In the multiplicity 1 case ($m = 1$) the proof now concludes easily. If

$$B(z) = \prod_{j=1}^{n} \frac{z - \alpha_j}{1 - \bar{\alpha}_j z} \tag{16.21}$$

then let

$$C_r(z) = \prod_{j=1}^{r} \frac{z - \alpha_j}{1 - \bar{\alpha}_j z}, \qquad 1 \leqslant r \leqslant n, \tag{16.22}$$

and let $C_0(z) \equiv 1$. The functions $\bar{C}_j f$, $0 \leqslant j \leqslant n$, are clearly linearly independent, they are $n + 1$ in number and, by the preceding paragraph, they are singular vectors of $H_{B\varphi}$. Their linear span F is an $(n+1)$-dimensional space satisfying (16.15). It follows that $n \leqslant k$ and thus that $\psi \in H_{(k)}^\infty$.

When s has multiplicity $m > 1$ there exist linearly independent singular vectors $f_1, \ldots, f_m$ of H_φ corresponding to s. The idea is that if we take $f = f_1$ and C_j, as above, then the functions

$$\bar{C}_0 f, \ldots, \bar{C}_n f, f_2, \ldots, f_m$$

will constitute $n + m$ linearly independent singular vectors of $H_{B\varphi}$ corresponding to s. However, to be sure of linear independence we must choose the right f_1. I assert that there is a singular vector f of H_φ such that $\bar{B}f$ is a rational function having no zeros in $\mathbb{D}$ (by (16.18), $\bar{B}f \in H^2$). To see this pick any singular vector h corresponding to s, and let $G = s^{-1}H_\varphi h$. By (16.18) we can write $h = Bh_0$ where $h_0 \in H^2$. If h_0 does not vanish on $\mathbb{D}$ we may take $f = h$. Otherwise, suppose that h_0 vanishes at $z_1, \ldots, z_r \in \mathbb{D}$, the zeros of the rational function h_0 in $\mathbb{D}$ being repeated according to their multiplicities. Then we may write $h_0 = vh_1$ where

$$v(z) = \prod_{j=1}^{r} \frac{z - z_j}{1 - \bar{z}_j z},$$

h_1 is a rational H^∞ function and h_1 has no zeros in $\mathbb{D}$. From (16.17),

$$\varphi h = \psi h + sG, \qquad \bar{\varphi}G = \psi^- G + sh,$$

which is to say

$$\varphi Bh_0 = \psi Bh_0 + sG, \qquad \bar{\varphi}G = \psi^- G + sBh_0.$$

Divide through by v to get

$$\varphi Bh_1 = \psi Bh_1 + s\bar{v}G, \qquad \bar{\varphi}\bar{v}G = \psi^- \bar{v}G + sBh_1.$$

Since $B\psi \in H^\infty$ and $h_1 \in H^2$, $\psi Bh_1 \in H^2$. Since $v \in H^\infty$ and $G \perp H^2$, $\bar{v}G \perp H^2$. Hence

$$H_\varphi(Bh_1) = P_-(\varphi Bh_1) = s\bar{v}G.$$

Further since $\psi^- G \perp H^2$, $\bar{v}\psi^- G \perp H^2$. Thus

$$H_\varphi^*(\bar{v}G) = P_+(\bar{\varphi}\bar{v}G) = sBh_1,$$

and so $\langle Bh_1, \bar{v}G \rangle$ is a Schmidt pair for H_φ corresponding to s. Let $f = Bh_1$: then $\bar{B}f\,(=h_1)$ is a rational H^2 function having no zeros in $\mathbb{D}$, and the assertion is justified.

Write

$$b_j(z) = \frac{z - \alpha_j}{1 - \bar{\alpha}_j z}$$

where the α_j are the zeros of B, so that $B = b_1 \ldots b_n$. Let $f = BF_1$ as in the last paragraph. By (16.20) the n functions

$$F_1, b_1 F_1, b_1 b_2 F_1, \ldots, b_1 b_2 \ldots b_{n-1} F_1 \tag{16.23}$$

are singular vectors of $H_{B\varphi}$ corresponding to s (if $C = b_{j+1} \ldots b_n$ then $\bar{C}f = b_1 \ldots b_j F_1$). Let X be the linear span of these vectors, and let Y be the space of singular vectors of H_φ corresponding to s (together with the zero vector). Now $X \cap Y = \{0\}$. For consider a singular vector f_2 of H_φ. By (16.18) we can write $f_2 = BF_2$ for some $F_2 \in H^2$. If $f_2 \in X$ we have, for some scalars λ_j,

$$\lambda_0 F_1 + \lambda_1 b_1 F_1 + \lambda_2 b_1 b_2 F_1 + \cdots + \lambda_{n-1} b_1 \ldots b_{n-1} F_1$$
$$= b_1 b_2 \ldots b_n F_2.$$

In this identity for rational H^2 functions put $z = \alpha_1$, so that $b_1(z) = 0$. By choice of f, $F_1(\alpha_1) \neq 0$, and so $\lambda_0 = 0$. Divide through by b_1 and then put $z = \alpha_2$ to get $\lambda_1 = 0$. Continuing in this way we deduce that $\lambda_0 = \cdots \lambda_{n-1} = 0$, so that $X \cap Y = \{0\}$ as claimed. The same argument, with F_2 replaced by 0, shows that the n vectors in (16.23) are linearly independent, so that dim $X = n$. Hence $X + Y$ is an $(n + m)$-dimensional subspace of H^2 consisting of singular vectors of $H_{B\varphi}$ corresponding to s. Retracing the argument we deduce that

$$s_{n+m-1}(H_\varphi) \geqslant s = s_{l+m-1}(H_\varphi) > s_{l+m}(H_\varphi) \geqslant \cdots$$

and so $n \leqslant l \leqslant k$ (cf. (16.14)). Hence $\psi \in H_{(k)}^\infty$.

We have now proved that the distance formula (16.2) is valid and that the distance is attained at a function $\psi \in H_{(k)}^\infty$ given by (16.3). We have still not quite finished, for our proof only showed that $\psi \in H_{(k)}^\infty$ for a particular choice of the singular vector f, whereas the theorem asserts we may use an arbitrary singular vector. It is conceivable that if we used a different f the resulting ψ might have too many poles. In fact, however, the same ψ results whichever f we use. Suppose f_1, f_2 are singular vectors of H_φ corresponding to s. Then

$$H_\varphi f_1 / f_1 = H_\varphi f_2 / f_2. \qquad (16.24)$$

To see this observe that, by Lemma 16.15,

$$(H_\varphi f_2)(H_\varphi f_2)^- = s^2 f_2 \bar{f}_2.$$

Hence

$$H_\varphi f_2 / f_2 = s^2 \bar{f}_2 / (H_\varphi f_2)^-.$$

Thus (16.24) is equivalent to

$$(H_\varphi f_1)(H_\varphi f_2)^- = s^2 f_1 \bar{f}_2.$$

This can be obtained by a slight modification of the proof of Lemma 16.15. The nth Fourier coefficient of the right hand side is $(s^2 f_1, z^n f_2)$ and reasoning almost identical to that in (16.12) shows that, for $n \geqslant 0$, this equals the nth Fourier coefficient of the left hand side. Interchanging f_1 and f_2 gives the same state for negative n, and so (16.24) is established.

It remains to prove the uniqueness assertion. We showed earlier in the proof that if the distance formula (16.2) is correct then any $\psi \in H_{(k)}^\infty$ at which the distance is attained satisfies

$$\psi = \varphi - (H_\varphi f)/f$$

for some singular vector f of H_φ corresponding to $s_k(H_\varphi)$. We now know that (16.2) *is* correct, the distance *is* attained and that the function $(H_\varphi f)/f$ is the same for all singular vectors f of H_φ corresponding to $s_k(H_\varphi)$. It follows that there is a unique best approximation $\psi \in H_{(k)}^\infty$ to φ with respect to the L^∞-norm. $\qquad \square$

At various points in the above proof we used the fact that an integrable function on $\partial \mathbb{D}$ is determined (up to sets of measure zero) by its Fourier coefficients. As we did not prove this theorem, it is worth pointing out that we can avoid it. We do know the result for continuous functions: it follows from Fejér's theorem. In the course of the proof it transpires that all the functions in question are continuous, being rational L^∞ functions on $\partial \mathbb{D}$. We could have arranged the proof (at some cost in intelligibility) to avoid appealing to the more general uniqueness result. It should be mentioned that the full Adamyan–Arov–Krein theorem is more general than what has been proved here. The distance formula (16.2) is true for *any* $\varphi \in L^\infty$, not just rational φ. The distance is always attained, though not always at a unique element of $H^\infty_{(k)}$. When H_φ has a singular vector corresponding to $s_k(H_\varphi)$ the uniqueness statement and the formula (16.3) are valid. In particular this is so when H_φ is compact, for example if φ is continuous. Only one new technique is needed to adapt the foregoing proof to the more general class of φs: arguments based on divisibility of polynomials have to be replaced by appeals to Beurling's theorem (see Hoffman, 1962). This states that all non-zero closed subspaces of H^2 which are invariant under multiplication by the independent variable have the form vH^2 for some inner function v. The theorem and its use are not especially difficult, but there is much to be said for understanding the rational case thoroughly first.

In Chapter 14 it was explained that in engineering applications there is a need for the extension of complex function theory to matrix-valued functions. Adamyan, Arov and Krein extended Theorems 15.14 and 15.16 to matrix-valued φ in 1971 and two engineers (S. Y. Kung and D. Lin) proved a version of Theorem 16.14 for matrix-valued φ in 1981. Elaboration of these results (for example, their use in numerical computation) and their application in engineering is the subject of intensive study.

16.4   Problems
16.1.   Let A be an $m \times n$ complex matrix and let the singular value decomposition of A be $A = UDV^*$ (so that U, V are unitary matrices and D is a non-negative diagonal matrix). What are the singular values and corresponding Schmidt pairs of the operator from $\mathbb{C}^n$ to $\mathbb{C}^m$ determined by A?

16.2. Let $A, B \in \mathscr{L}(H, K)$, where H, K are Hilbert spaces, and let $A^*A \leqslant B^*B$. Show that $s_k(A) \leqslant s_k(B)$ for any non-negative integer k.

16.3. Assuming the existence of discontinuous linear functionals, prove that not every subspace of finite codimension in a Hilbert space is closed.

16.4. Let s be a singular value of multiplicity m of an operator T. Show that if $s > 0$ then s is a singular value of T^* of multiplicity m. Give an example to show that, when $s = 0$, s need not be a singular value of T^*.

16.5. Is $H^\infty_{(k)}$ a closed subset of L^∞?

16.6. Let B, C be inner functions. Prove that $H^2 \ominus (BC)H^2$ is the orthogonal direct sum of $H^2 \ominus BH^2$ and $B(H^2 \ominus CH^2)$. Hence give an alternative proof that if B is a finite Blaschke product of degree k then BH^2 has codimension k in H^2.

16.7. Show that, for $g \in L^2$, $(P_+ g)^- = zP_-(\bar{z}\bar{g})$ and $(P_- g)^- = zP_+(\bar{z}\bar{g})$.

16.8. Let $\varphi \in L^\infty$ and let g be a singular vector of H_φ corresponding to a singular value s. Prove that zg is a singular vector of $H_{z\varphi}$ corresponding to s.

16.9. Let

$$\varphi(e^{i\theta}) = 1 + 2\cos\theta + 2\cos 2\theta, \qquad 0 < \theta < 2\pi.$$

Show that

$$\text{dist}_{L^\infty}(\varphi, H_{(1)}^\infty) = (\sqrt{5} - 1)/2,$$

and find the best approximation to φ in $H_{(1)}^\infty$ with respect to the L^∞-norm.

16.10. Let $\alpha \in D$ and let

$$\varphi(z) = \frac{1}{z(z - \alpha)}, \qquad z \in D.$$

Find $\text{dist}_{L^\infty}(\varphi, H_{(1)}^\infty)$ and find the best approximation to φ in $H_{(1)}^\infty$ with respect to the L^∞-norm.

16.11. Let T_1, T_2 be as in Problem 15.12. Show that

$$\inf \| T_1 + T_2 Q \| = \sqrt{3} - 1$$

where Q ranges over $H_{(1)}^\infty$, and find the unique function $Q \in H_{(1)}^\infty$ for which the infimum is attained.

16.12. Show that the infimum of the L^∞-norms of all functions $f \in H_{(1)}^\infty$ which satisfy

$$f(0) = 1, \qquad f(\tfrac{1}{2}) = 2$$

is $\sqrt{3} - 1$, and find the unique interpolating function $f \in H_{(1)}^\infty$ of minimal norm.

APPENDIX: SQUARE ROOTS OF POSITIVE OPERATORS

Proof of the existence of square roots of positive operators was deferred in Chapter 12 in the interest of smoother exposition: this was justified by the fact that the desired application required the existence only for diagonalizable positive operators, in which case it is obvious. However, to show that the theorems are all valid in the stated generality we do need to fill the gap. Throughout this appendix H denotes an arbitrary Hilbert space.

Lemma 1 Let $A \in \mathcal{L}(H)$ be positive. For any $f \in H$,
$$\|Af\|^2 \leqslant \|A\|(Af, f).$$
Proof. The quadratic function
$$(A(f + tAf), f + tAf) = (Af, f) + 2\|Af\|^2 t + (A^2 f, Af) t^2$$
is non-negative for all $t \in \mathbb{R}$. It therefore has a non-positive discriminant: that is,
$$\begin{aligned}
\|Af\|^4 &\leqslant (Af, f)(A^2 f, f) \\
&\leqslant (Af, f)\|A^2 f\|\|Af\| \\
&\leqslant \|A\|(Af, f)\|Af\|^2.
\end{aligned}$$
$\qquad\qquad\square$

Lemma 2 Let $(A_n)_1^\infty$ be a sequence in $\mathcal{L}(H)$ such that
$$0 \leqslant A_1 \leqslant A_2 \leqslant \cdots \leqslant I.$$
Then there exists $A \in \mathcal{L}(H)$ such that
$$\lim_{n \to \infty} A_n f = Af$$
for all $f \in H$.

Proof. Putting $A = A_n - A_m$ in Lemma 1 yields

$$\|A_n f - A_m f\|^2 \leqslant \|A_n - A_m\|((A_n - A_m)f, f)$$
$$\leqslant ((A_n - A_m)f, f).$$

Since $(A_n f, f)$, $n \in \mathbb{N}$, is a non-decreasing bounded sequence in $\mathbb{R}$ it is Cauchy. Hence $(A_n f)_{n \in \mathbb{N}}$ is Cauchy in H, and so it has a limit in H which we may denote by Af. This defines A as a bounded linear operator on H. □

Theorem Let A be a positive operator on a Hilbert space H. There exists a unique square root $A^{1/2}$ of A in $\mathscr{L}(H)$. That is, there exists a unique positive operator $B \in \mathscr{L}(H)$ such that $B^2 = A$. Moreover there exists a sequence (p_n) of polynomials such that

$$p_n(A)f \to A^{1/2}f \qquad \text{as } n \to \infty$$

whenever $f \in H$ and A is a positive contraction.

Proof. We shall first prove the existence statement by an iterative scheme. We can suppose $0 \leqslant A \leqslant I$. We seek X, $0 \leqslant X \leqslant I$, such that $X^2 = A$. Put $B = I - A$, $Y = I - X$. The desired Y will satisfy $(I - Y)^2 = I - B$, that is,

$$Y = \tfrac{1}{2}(B + Y^2). \tag{A1}$$

Conversely, if $0 \leqslant Y \leqslant I$ and (A1) holds then the operator $X = I - Y$ is a positive square root of A. We solve (A1) by successive approximation: let

$$Y_0 = 0, \qquad Y_{n+1} = \tfrac{1}{2}(B + Y_n^2), \qquad n = 0, 1, 2, \ldots .$$

By induction, $Y_n \geqslant 0$ for all $n \in \mathbb{N}$. Since

$$I - Y_{n+1} = \tfrac{1}{2}A + \tfrac{1}{2}(I - Y_n^2)$$

it also follows by induction that $Y_n \leqslant I$ (which is equivalent to $\|Y_n\| \leqslant 1$, by Theorem 7.18). It is also true that $Y_n - Y_{n-1} \geqslant 0$ for all $n \in \mathbb{N}$: this is slightly more subtle. We have

$$Y_{n+1} - Y_n = \tfrac{1}{2}(B + Y_n^2) - \tfrac{1}{2}(B + Y_{n-1}^2)$$
$$= \tfrac{1}{2}(Y_n^2 - Y_{n-1}^2).$$

Since Y_n and Y_{n-1} are both polynomials in B (again by induction) they commute with one another, and so

$$Y_{n+1} - Y_n = \tfrac{1}{2}(Y_n + Y_{n-1})(Y_n - Y_{n-1}),$$

for $n \in \mathbb{N}$, while $Y_1 - Y_0 = \tfrac{1}{2}B$. Let us define sequences $(q_n), (r_n)$ of polynomials in a complex variable b by

$$q_0(b) = 0, \qquad r_0(b) = \tfrac{1}{2}b,$$
$$q_{n+1}(b) = \tfrac{1}{2}(b + q_n(b)^2),$$
$$r_{n+1}(b) = \tfrac{1}{2}(q_{n+1}(b) + q_n(b))r_n(b).$$

Then it is clear by induction that q_n, r_n are polynomials in b having all coefficients non-negative. Further,

$$Y_n = q_n(B), \qquad Y_{n+1} - Y_n = r_n(B),$$

$n = 0, 1, 2, \ldots$. Thus $Y_{n+1} - Y_n$ is a sum of non-negative multiples of non-negative powers of the positive operator B, and so $Y_{n+1} - Y_n \geqslant 0$ (recall Problem 12.5).

By Lemma 2 there exists $Y \in \mathscr{L}(H)$ such that $Y_n f \to Yf$ for all $f \in H$. Then

$$\begin{aligned}
(Yf, f) &= \lim(Y_{n+1} f, f) \\
&= \lim \tfrac{1}{2}((B + Y_n^2)f, f) \\
&= \tfrac{1}{2}(Bf, f) + \lim \|Y_n f\|^2 \\
&= \tfrac{1}{2}((B + Y^2)f, f).
\end{aligned}$$

By polarization (cf. Problem 12.3) it follows that (A1) holds. Clearly $0 \leqslant Y \leqslant I$, and so $X = I - Y$ is a positive square root of A.

Now let us construct the polynomials p_n. We have, for any $f \in H$,

$$\begin{aligned}
Yf &= \lim Y_n f = \lim q_n(B)f \\
&= \lim q_n(I - A)f,
\end{aligned}$$

so that

$$Xf = \lim\{I - q_n(I - A)\}f.$$

It therefore suffices to take $p_n(a) = 1 - q_n(1 - a)$, to give

$$p_n(A)f \to Xf \qquad \text{as } n \to \infty \tag{A2}$$

for all $f \in H$.

We have given a procedure which assigns to every positive contraction A a positive square root X such that (A2) holds, (p_n) being a fixed sequence of complex polynomials. Let us temporarily denote this square root by $A_{1/2}$, the unorthodox subscript being to emphasize that, as far as we know so far, $A_{1/2}$ may be only one of many positive square roots of A. In fact it is not. Suppose B is any positive square root of A: we shall show that $B = A_{1/2}$.

B commutes with A, since

$$BA = B^3 = AB.$$

Hence B commutes with any power of A, and so with any polynomial in A. Thus

$$Bp_n(A)f = p_n(A)Bf$$

for any $n \in \mathbb{N}$ and $f \in H$. Letting $n \to \infty$ and using (A2) we obtain

$$BA_{1/2} = A_{1/2}B. \tag{A3}$$

Now let

$$M = (A_{1/2})_{1/2}, \qquad N = B_{1/2}.$$

Pick $f \in H$ and let $g = (A_{1/2} - B)f$. We have

$$\|Mg\|^2 + \|Ng\|^2 = (M^2g, g) + (N^2g, g)$$
$$= ((A_{1/2} + B)g, g)$$
$$= ((A_{1/2} + B)(A_{1/2} - B)f, g).$$

By (A3),

$$(A_{1/2} + B)(A_{1/2} - B) = A_{1/2}^2 - B^2 = A - A = 0.$$

Thus $Mg = 0 = Ng$, and therefore $M^2g = 0 = N^2g$, i.e. $A_{1/2}g = 0 = Bg$. It follows that

$$\|(A_{1/2} - B)f\|^2 = ((A_{1/2} - B)^2f, f)$$
$$= ((A_{1/2} - B)g, f)$$
$$= 0.$$

That is, $A_{1/2} = B$, and so there is only one positive square root of A. $\quad\square$

REFERENCES

de Barra, G. (1981) *Measure Theory and Integration*, Ellis Horwood.

Binmore, K. G. (1981) *Foundations of Analysis: A Straightforward Introduction*, Book 1: *Logic, Sets and Numbers;* Book 2: *Topological Ideas*, Cambridge University Press.

Dieudonné, J. (1981) *History of Functional Analysis*, North Holland.

Dunford, N. and Schwartz, J. T. (1957) *Linear Operators Part 1*, Interscience.

Dunford, N. and Schwartz, J. T. (1963) *Linear Operators Part 2*, Interscience.

Francis, B. A. (1986) *A Course in H^∞ Control Theory*, Springer Verlag.

Friedman, A. (1969) *Partial Differential Equations*, Holt, Rinehart and Winston.

Hardy, G. H. (1940) *A Mathematician's Apology*, Cambridge University Press (second edition 1967).

Hardy, G. H. and Rogosinski, W. W. (1965) *Trigonometric Series* (third edition), Cambridge University Press.

Helson, H. (1964) *Lectures on Invariant Subspaces*, Academic Press.

Hoffman, K. (1962) *Banach Spaces of Analytic Functions*, Prentice-Hall.

Nikolskii, N. K. (1985) *Treatise on the Shift Operator*, Springer Verlag.

Reid, C. (1970) *Hilbert*, Allen and Unwin.

Rudin, W. (1966) *Real and Complex Analysis*, McGraw-Hill.

Rudin, W. (1973) *Functional Analysis*, McGraw-Hill.

Simmons, G. F. (1972) *Differential Equations, with Applications and Historical Notes*, McGraw-Hill.

Sneddon, I. N. (1961) *Special Functions of Mathematical Physics and Chemistry* (second edition), Oliver and Boyd.

Titchmarsh, E. C. (1962) *Eigenfunction Expansions* (second edition), Oxford University Press.

Tricomi, F. G. (1961) translated by E. A. McHarg, *Differential Equations*, Blackie.

Watson, G. N. (1944) *A Treatise on the Theory of Bessel Functions* (second edition), Cambridge University Press.

1.1. $\operatorname{Re} s > \frac{1}{2}$.

1.7. $(f, g) = \displaystyle\int_{-1}^{1} |x| f(x)\overline{g(x)} + 3f'(x)\overline{g'(x)}\, dx$.

1.9. Apply Cauchy–Schwarz to the functions x^2, f in the inner product space of Problem 1.7.

1.10. Let $\omega = e^{2\pi i/k}$. Then

$$(x, y) = \frac{1}{k}\sum_{n=0}^{k-1}\omega^n\|x + \omega^n y\|^2.$$

2.6. exp is not a polynomial but belongs to the closure of the polynomials with respect to either norm.

2.12. Let V be the inner product space $C[0,1]$ of Exercise 1.3, $E = \{f \in V: f(0) = 0\}$.

3.5. The partial sums of the Maclaurin series for exp are Cauchy but not convergent in RH^2.

3.6. $f_\alpha \in L^2(0, \infty) \Leftrightarrow \alpha > -\frac{1}{2}$.

$$\|f_\alpha\| = 2^{-\alpha-1/2}\{\Gamma(2\alpha + 1)\}^{1/2}, \qquad \alpha > -\tfrac{1}{2}.$$

4.1. $(1, \omega^2, \omega)$.

4.5. If $|\alpha| < 1$ the nth Fourier coefficient is 0 if $n \geq 0$, α^{-n-1} if $n < 0$. If $|\alpha| > 1$ it is 0 if $n < 0$, $-\alpha^{-n-1}$ if $n \geq 0$.

4.8. $\frac{2}{3}(1, -2, 1)$.

4.9. $8/175$.

4.16. Yes.

4.18. $\mathbb{C}y$.

6.1. It does not make sense, since elements of L^2 are only defined almost everywhere.

6.4. $\sqrt{n}$.

6.5. $\|F\| = \frac{1}{2}$. F attains its norm at the constant function equal to 1.

7.1. $\|D\| = \sup_n |\lambda_n|$.

7.5. No.

7.12. $\|D^{-1}\| = (\inf_n |\lambda_n|)^{-1}$ when this exists.

7.15. $f(t) = \sin t + \displaystyle\int_0^t e^{t-s} \sin s \, ds$.

7.17. No: think of the forward and backward shifts on ℓ^2.

7.22. $\frac{1}{2}(3 + \sqrt{5})$.

7.23. $\sqrt{\{\frac{1}{2}(3 + \sqrt{5})\}}$.

7.24. $T^*X = U^*XV^*$.

7.26. $T^*x = (x, \psi)\varphi$.

7.34. $\mathrm{clos}\{\lambda_n : n \in \mathbb{N}\}$.

7.36. $\{0, (\psi, \varphi)\}$ assuming $\dim H > 1$.

7.38. The closed unit disc.

8.2. (a), (b) are false, (c) is true.

8.7. Neither.

8.8. No: consider $f_n(x) = n^{-1} e^{inx}$, $n \in \mathbb{N}$.

9.6. (a) Eigenvalues $\{(k + \frac{1}{2})\pi\}^2$, $k = 0, 1, 2, \ldots$. Corresponding eigenfunctions $\cos(k + \frac{1}{2})\pi x$.

(b) Eigenvalues α_k^2, where $\alpha_1, \alpha_2, \ldots$ are the positive solutions of $\tan \alpha = -\frac{1}{3}\alpha$. Graph indicates that $\alpha_k \sim (k - \frac{1}{2})\pi$ for large k. Corresponding eigenfunction is $\sin \alpha_k x$.

(c) Eigenvalues $k^2\pi^2 + \frac{1}{4}$, $k \in \mathbb{N}$. Corresponding eigenfunction is $e^{-\frac{1}{2}x} \sin k\pi x$.

(d) If $F(t) = f(e^t)$, equation is

$$F'' + 2F' + \lambda F = 0, \qquad F(0) = 0 = F(1).$$

Eigenvalues are $k^2\pi^2 + 1$, $k \in \mathbb{N}$, corresponding eigenfunctions $F(t) = e^{-t} \sin k\pi t$, i.e. $f(x) = x^{-1} \sin(k\pi \log x)$.

10.2. Green's function is

$$k(x, \tau) = \frac{1}{\omega \sin \omega} \times \begin{cases} \cos \omega(x - 1) \cos \omega\tau & \text{if } 0 \leqslant \tau \leqslant x \leqslant 1, \\ \cos \omega x \cos \omega(\tau - 1) & 0 \leqslant x \leqslant \tau \leqslant 1. \end{cases}$$

$\omega = 0$ is an eigenvalue, so no Green's function in this case.

10.3. $k(x, \tau) = \begin{cases} x - 2 & 0 \leqslant \tau \leqslant x \leqslant 1, \\ \tau - 2 & 0 \leqslant x \leqslant \tau \leqslant 1. \end{cases}$

12.2. K is positive in case (a), not in (b) or (c).

12.10. Use

$$\begin{bmatrix} A & B \\ B^* & C \end{bmatrix} = \begin{bmatrix} I & 0 \\ B^*A^{-1} & I \end{bmatrix} \begin{bmatrix} A & 0 \\ 0 & C - B^*A^{-1}B \end{bmatrix} \begin{bmatrix} I & A^{-1}B \\ 0 & I \end{bmatrix}.$$

12.11. $\begin{bmatrix} A & I \\ I & A^{-1} \end{bmatrix}^{1/2} = \begin{bmatrix} (A+A^{-1})^{-1/2}A & (A+A^{-1})^{-1/2} \\ (A+A^{-1})^{-1/2} & (A+A^{-1})^{-1/2}A^{-1} \end{bmatrix}.$

12.15. $\frac{1}{2}(1+\sqrt{17})$; $a = \frac{1}{4}(1-\sqrt{17})$.

15.3. $(P_+\varphi)(e^{it}) = -\frac{2i}{\pi}\left(\frac{e^{it}}{1} + \frac{e^{i3t}}{3} + \frac{e^{i5t}}{5} + \cdots\right)$

$= \begin{cases} \dfrac{1}{2} + \dfrac{i}{\pi}\log\tan\dfrac{t}{2}, & 0 < t < \pi, \\[2mm] -\dfrac{1}{2} + \dfrac{i}{\pi}\log\left|\tan\dfrac{t}{2}\right|, & -\pi < t < 0. \end{cases}$

15.7. $1 + e^{i\theta} + e^{i2\theta} + \dfrac{\frac{1}{2}(\sqrt{5}-1)}{1 + \frac{1}{2}(\sqrt{5}-1)e^{i\theta}}.$

15.9. Distance $= \dfrac{|\alpha| + \sqrt{(4-3|\alpha|^2)}}{2(1-|\alpha|^2)},$

$g(z) = \dfrac{\bar{\alpha}^2(|\alpha| + \sqrt{(4-3|\alpha|^2)})^2}{2(1-|\alpha|^2)\{|\alpha|(|\alpha| + \sqrt{(4-3|\alpha|^2)}) + 2(1-|\alpha|^2)\bar{\alpha}z\}}.$

15.12. $Q(z) = \dfrac{1}{1-\frac{1}{2}z}\left(1 - \dfrac{2\sqrt{3}}{(\sqrt{3}-1)z+2}\right).$

15.13. $f(z) = 1 + 2z + \dfrac{z(z-\frac{1}{2})}{1-\frac{1}{2}z}\left(1 - \dfrac{2\sqrt{3}}{(\sqrt{3}-1)z+2}\right)$

$= (\sqrt{3}+1)\dfrac{z + \dfrac{\sqrt{3}-1}{2}}{1 + \dfrac{\sqrt{3}-1}{2}z}.$

16.1. The jth diagonal entry of D is a singular value of A and the jth columns of V, U respectively make up a corresponding Schmidt pair.

16.4. Take $T =$ backward shift on ℓ^2.

16.5. Yes.

16.10. $1 + e^{i\theta} + e^{i2\theta} - \dfrac{\frac{1}{2}(\sqrt{5}+1)}{1 - \frac{1}{2}(\sqrt{5}+1)e^{i\theta}}.$

16.11. Distance $= \dfrac{|\alpha| - \sqrt{(4-3|\alpha|^2)}}{2(1-|\alpha|^2)},$

$\psi(z) = \dfrac{\bar{\alpha}^2(|\alpha| - \sqrt{(4-3|\alpha|^2)})^2}{2(1-|\alpha|^2)\{|\alpha|(|\alpha| - \sqrt{(4-3|\alpha|^2)}) + 2(1-|\alpha|^2)\bar{\alpha}z\}}.$

16.12. $Q(z) = \dfrac{1}{1-\frac{1}{2}z}\left(1 - \dfrac{2\sqrt{3}}{(\sqrt{3}+1)z-2}\right).$

16.13. $f(z) = 1 + 2z + \dfrac{z(z - \frac{1}{2})}{1 - \frac{1}{2}z}\left(1 - \dfrac{2\sqrt{3}}{(\sqrt{3} + 1)z - 2}\right)$

$= -(\sqrt{3} - 1)\dfrac{z - \dfrac{\sqrt{3} + 1}{2}}{1 - \dfrac{\sqrt{3} + 1}{2}z}.$

What prospects are in view, what further peaks accessible from the high ground of Hilbert space? We trust our teachers, that the agony of learning they inflict on us is to good purpose, but resolution flags, and may perhaps be quickened by the contemplation of our goal.

Euclid was able to permit himself a lordly response to a pupil who asked about the uses of geometry. According to tradition Euclid summoned a servant and instructed him to give a penny to the pupil so that he should have profit from his studies. If your sympathies are with the recipient of the coin try reading *A Mathematician's Apology* (Hardy, 1940) for an updated and cogently argued presentation of the Euclidean attitude. Hardy firmly divides mathematics into two kinds: the useful but boring and 'real mathematics' which is beautiful but of no use whatever. In his inaugural lecture as professor at Oxford Hardy came close to saying that the value of mathematics is to absorb harmlessly the intellectual energies of very clever people who might otherwise be making weapons of destruction. Hardy was the most influential British mathematician of the day and his views took a long time to live down. Since then higher education has grown hugely, more colleges and new disciplines are vying for their share of the public purse, and we are all more conscious of the scrutiny of the Treasury and more careful of the public's perception. Hardy sought to justify doing mathematics from the standpoint of the individual, but surely it is right that we should be willing to answer to society for our activities. This does not mean we must be crudely utilitarian, but it does mean we should be ready to give an account of the part played by mathematics as a whole in science, technology, industry, commerce and government, and that we should be subject to political judgement as to the resources it deserves. The community may decide it needs comparatively few scholars who can write, like Hardy,

I have never done anything 'useful'. No discovery of mine has made, or is likely to make, directly or indirectly, for good or ill, the least difference to the amenity of the world (Hardy, 1940, §29).

Must we accept Hardy's dichotomy as the truth? It would be distressing were it so. We look round the universities, research institutions and companies and see wonderful advances being made in the life sciences, physics, chemistry and engineering which will have a great impact on society. If 'real mathematics' is an island for ever cut off from interaction with the mainland of science and technology then many of us would wish to devote our talents elsewhere. The beauty and profundity of mathematics are undeniable, and they have an aesthetic value greatly exceeding that of many works supported in one way or another by our society, but nevertheless I would make the case for the present level of research activity in mathematics as part of the case for science and technology as a whole, and that is certainly not an entirely aesthetic one.

I believe there are many grounds for rejecting the dichotomy. In the first place there is much less agreement among mathematicians than Hardy implies as to what is boring, what is beautiful and what is 'real mathematics'. Nor need we accept the claim that it is only those parts of mathematics which are commonplace and dull that count for practical life (Hardy, 1940, §25). It was not so in Newton's time, and even if it had become true by 1940 it is false again half a century later. There are many areas of fundamental research which relate directly to scientific or technological questions. One example is the theory of solitons, which looks like being practically important; another is the problem of robustness in control which was briefly described in Chapter 14. It is significant that in this connection engineers routinely use the spaces of functions introduced by Hardy himself. It certainly looks as though Hardy's assessment in the paragraph quoted above was mistaken.

But there is a more fundamental objection to Hardy's view of science (in which I include for the moment mathematics and engineering): it overemphasizes the individual. Therein it reflects a liberal political philosophy of the beginning of the century. Changes in intellectual fashion have made it easier for us to see science as a social enterprise. Science is not just a collection of facts and theories nor of books and journals: the living body of science subsists in the learning, research and interaction of its practitioners. It is a grand structure of which no one person can see more than a tiny part. Even if I were to restrict myself to mathematics journals, I could understand much less than 1% of the hundreds of thousands of papers which appear every year. Nevertheless, the particular combination of things I know and connections I can make is unique to me, and by reading, discovering, writing, talking, lecturing and, if I work in industry, helping to make something, I play my small part in keeping science alive, like a cell in a body. Science as a whole is abundantly fruitful, in many, diverse ways. Mathematicians, in the legitimate pursuit of beauty and understanding, have served their colleagues in science well. They have provided effective conceptual tools, and where something new was called for they have been willing to tackle new mathematical problems. Some may only be interested in cultivating the brightest blooms in the garden of mathematics, but many others are keeping in good order the great variety of plants asked for by those who need to enter, and even the bright blooms surprisingly often bear edible fruit.

Why this philosophical disquisition in a book on Hilbert space? In lectures and tutorials we are so preoccupied with proving theorems and solving problems that we rarely stand back and survey mathematics in society, and yet it is important for students to form a view on this matter, as it will influence their attitude to study, their choice of subjects and indeed of career. I have often found students to have a view of the subject so distorted as to impede their development as scientists. Intellectual snobbery consisting of contempt for applications is quite common: this is usually transient, but is harmful while it lasts. From the viewpoint I have outlined I shall say a little about the place of functional analysis in mathematics courses, in the individual's learning of mathematics and in mathematics itself as a branch of science.

Functional analysis has certainly been one of the major strands of development in twentieth-century mathematics. A very large number of research papers are identifiably in this area, many working mathematicians regard themselves as specialists in it and it has attracted some of the greatest mathematicians (e.g. Hilbert, F. Riesz, von Neumann, S. Banach, I. M. Gelfand, M. G. Krein, A. Grothendieck and A. Connes). It is not the only part of analysis which has thrived: among other topics, trigonometric series and complex analysis have been much pursued. These theories, however, were at a different stage of their development. By 1900 they were mature theories, known to mathematicians generally and already interacting with other branches of science. Functional analysis, on the other hand, was just being born. Some strands in its lineage have been traced in earlier chapters but, irrespective of how it came about, it is clear from a present perspective that it was an inevitable development: mathematicians on other planets surely also know about orthogonal expansions and eigenvalues of compact Hermitian operators. It constitutes a fusion of basic algebraic and analytic structures which was necessary for progress over a broad front. For example, the idea of diagonalization, which is one of the most pervasive and powerful techniques in mathematics, occurs in many guises and cries out for a framework of infinite-dimensional linear spaces.

So the time was ripe for functional analysis, F. Riesz and others gave a glimpse of the power and elegance waiting to be revealed, governments were beginning to invest more heavily in basic science and we flocked in, great and small, and built a splendid edifice. Like other great structures of mathematics it has many and various high pinnacles, of which neighbouring citizens have only a hazy impression, but its bulk has become part of the landscape, to which all may refer with the expectation of recognition. Workers in the traditional areas of applied mathematics (the ones more or less closely related to physics) would practically all be familiar with the contents of Chapters 1 to 8, and with quite a lot more which is not in this book, such as the elements of the theory of distributions (generalized functions). Many theoretical engineers also regularly use this material. Differential topologists use Fredholm operators, which are just beyond the scope of this book, and are starting to invoke more advanced parts of the theory of algebras of operators. Numerical analysis has need of the concepts and methods of functional analysis. A mathematician or engineer working in industry who uses eigenfunction expansions as a standard technique would very likely *not* know in detail the functional analytic background: there would doubtless be too many other things more important for him to know. However, if he wished to innovate and refine on the technique he might well need to have recourse to an individual or book who could provide the necessary piece of operator theory.

The great majority of applications of functional analysis are drawn from the rudiments of the theory, but not all are, and no one can tell what topics will become important. Chapter 14 provides an example: engineering theorists in a certain area have found some comparatively advanced operator theory relevant to their needs.

Should we infer that every degree course in mathematics needs to include some functional analysis? Much as I should like this book to sell many copies, I would not make such a claim. There are topics which every mathematician should know: I would include, for example, the basics of matrices, Fourier series, elementary differential equations and complex analysis. However, I have already implied my view that the vitality of science is enhanced by its practitioners' having a wide spread of types of knowledge. Nowadays there are many sorts of mathematics degree, some oriented towards management, operations research or mathematical modelling. It would be rash of me to try and prescribe for them. But for those courses in the high tradition of mathematics, which aspire to bring students within reach of the frontiers of mainstream research, I believe there should be at least an option of studying basic Hilbert space theory.

What of the present and future of research in functional analysis? It is hard to say much of use about so vast and problematic a question in a few words, but the student who is contemplating a career in mathematical research needs what guidance he can get. I will risk outlining some of my own preferences and prognostications, trusting the student to seek as many other opinions as he can find.

I have written as if functional analysis were a single well-defined topic, but in reality its boundaries are vague and it includes several distinct branches, which can prosper or decline almost independently. For a couple of decades the theory of locally convex topological vector spaces had a high vogue. It originated from the study of distributions, unquestionably a fundamental notion, was seen as an important topic and inspired deep researches and beautiful theorems. But quite suddenly in the early 1970s it ran out of momentum. Its results are still used, there remain hard problems and some people continue to work in it. But in the main mathematicians moved out into adjoining areas. The theory seemed to have gone as far as it usefully could for the time being: other topics appeared to offer a better return on effort.

So there is a little of a gamble involved in selecting a PhD subject: no one wants to board a ship which is about to run aground. Are any other vessels in the fleet approaching shoals? It could be one which is steaming ahead at top speed. The structure theory of Banach spaces is at present one of the principal foci of activity in functional analysis; it has enjoyed the attention of some outstanding talents for around twenty-five years and stands high in prestige. Could it be that the problems which are both worthwhile and accessible are nearly all solved and it in turn is due for a fallow period? This is sheer speculation: I cannot pretend to either the breadth of knowledge or the courage to make a prediction. It could well be that it will continue to thrive by virtue of its connections with probability, complex analysis and combinatorics.

A subject which *is* probably past its heyday (again I qualify – for the time being!) is the general structure theory of Banach algebras (Banach spaces which are also rings with respect to a continuous multiplication). These are important entities and their theory is particularly elegant. It continues to have a strong influence throughout analysis, but my impression is that mathematicians are choosing to give it a rest after around forty-five years of intensive study. They are carrying its insights into more specialized structures.

One substantial area which is definitely waxing rather than waning is the study of operator algebras (Banach algebras whose elements are operators on a Hilbert space). This was initiated by von Neumann in the 1930s and 1940s, and his fame attracted good mathematicians into the field. It made steady progress and had many successes over the years, but was possibly somewhat inward looking. Recently, however, it has been transformed by the genius of the Frenchman Alain Connes. In addition to making great advances in the theory itself he has brought it into contact with algebraic topology and differential geometry in a synthesis which looks like being extremely fruitful. It is a hard subject to profess because of the high level of technique it requires and the very rapidity of its growth, but it offers many opportunities for an able and ambitious student.

'Operator theory' usually means the study of individual operators, or classes of operators with some common structure, rather than algebras of operators. The two branches do of course interact, but they have quite different flavours. Operator theory has also been well worked ever since Hilbert, F. Riesz and von Neumann, and is now an edifice of great extent and richness with many active specialists.

So far everything I have said has related to *linear* analysis. For a long time scientists concentrated mainly on linear models of natural phenomena (laws describable by linear differential equations). These are often satisfactory, and the difficulty of non-linear models looks daunting. However, the nettle must be grasped: some phenomena are irredeemably non-linear (for example, solitary waves). There is much activity in non-linear functional analysis: it is clearly a theory with a long way to go.

I will conclude with two pieces of advice for would-be analysts. First, tackle a problem whose solution will be of interest to a goodly number of people: do not seek refuge in a backwater. One expedient would be to choose a famous outstanding problem, but this is a high-risk policy. Better (though still not a policy for everyone) is to go for something with at least an applied slant. At any rate, at this stage of your career, try to find an agreeable tide and swim with it.

My parting advice is a thought for the longer term. Many of the branches within functional analysis seem to me to have reached such a high state of development that outward expansion has become more of a priority. The most significant work in the near future may come from uniting functional analysis with other branches of mathematics and science. As I have said, this is already happening in operator algebras. It is perhaps being rather slow to happen in operator theory. While you are serving your apprenticeship as a research worker you will probably find it hard enough to penetrate to a sufficient depth in a single area, but once you have succeeded, I counsel investing in breadth.

SUBJECT INDEX